Translations of
MATHEMATICAL MONOGRAPHS

Volume 223

Painlevé Equations through Symmetry

Masatoshi Noumi

Translated by
Masatoshi Noumi

American Mathematical Society
Providence, Rhode Island

パンルヴェ方程式
-対称性からの入門-

PANRUVE HOTEISHIKI
by Masatoshi Noumi

Translated from the Japanese by the author.

2000 *Mathematics Subject Classification.* Primary 34M55;
Secondary 37K35, 37K10, 39Axx, 14E05, 20F55.

For additional information and updates on this book, visit
www.ams.org/bookpages/mmono-223

Library of Congress Cataloging-in-Publication Data

Noumi, Masatoshi, 1955–
[Panruve hoteishiki. English]
Painlevé equations through symmetry / Masatoshi Noumi ; translated by Masatoshi Noumi.
p. cm. — (Translations of mathematical monographs, ISSN 0065-9282 ; v. 223)
Includes bibliographical references and index.
ISBN 0-8218-3221-2 (acid-free paper)
1. Painlevé equations. 2. Bäcklund transformations. 3. Hamiltonian systems. I. Title. II. Series.

QA372.N66713 2004
515′.352—dc22 2003062828

10 9 8 7 6 5 4 3 2 1   09 08 07 06 05 04

Contents

Preface

The title of this book, *Painlevé equations*, refers to a set of six nonlinear ordinary differential equations denoted traditionally by $P_{\rm I}$, $P_{\rm II}$, ..., $P_{\rm VI}$. The first one

$$P_{\rm I}: \quad y'' = 6y^2 + t$$

appears to be the simplest among the six. Here t stands for the independent variable, $y = y(t)$ for the dependent variable (or the unknown function), and $' = d/dt$ for the derivative with respect to t. This is a *nonlinear* differential equation since its right-hand side is a quadratic function of y. As you go on to $P_{\rm II}$, $P_{\rm III}$, ..., the Painlevé equations take on a far more complicated appearance. I think you might be surprised if you are seeing $P_{\rm VI}$ for the first time! (See Table 0.1.)

The six Painlevé equations were discovered about a hundred years ago by a French mathematician named Paul Painlevé (1863–1933). Their solutions are now called the *Painlevé transcendents.* (Incidentally Painlevé was a politician too; you could see a photo of Painlevé celebrating Lindbergh, the first person to fly an airplane from New York to Paris.) You might ask "Could there be anything left unstudied about differential equations that have been known for a hundred years?" I would say, "Yes, they are *special.*" In fact the history of the Painlevé equations reads like a detective story. But the Painlevé equations themselves are really a wonder. They still continue to give us fresh mysteries. You may suspect I am exaggerating, but this is my true impression. One reason I wrote this book is to tell you how impressed I am by the mysteries of the Painlevé equations.

The main theme of this book is the *symmetry* of the Painlevé equations. I will focus on $P_{\rm II}$ and $P_{\rm IV}$, namely,

$$\begin{aligned} P_{\rm II}: &\qquad y'' = 2y^3 + ty + \alpha, \\ P_{\rm IV}: &\qquad y'' = \frac{1}{2y}(y')^2 + \frac{3}{2}y^3 + 4ty^2 + 2(t^2 - \alpha)y + \frac{\beta}{y}, \end{aligned}$$

as examples, since they are easier to handle than the others. The symbols α and β appearing in these equations are parameters. If you are given a solution of a Painlevé equation with some parameters, you can use a special device, called a *Bäcklund transformation*, for creating a new solution with different parameters. "Symmetry" is a word used frequently to refer to such a mechanism to construct new solutions by transformations. I will begin by explaining what a Bäcklund transformation is and tell the story of the symmetry of the Painlevé equations on the basis of explicit examples. I hope you enjoy seeing how new solutions are generated, one after another, starting from simple solutions. I will also introduce the τ-functions (tau functions) and describe symmetry as a discrete system with respect to the parameters. After that I will formulate (in Chapter 5) the *Jacobi-Trudi formula*, a way to express explicitly all the solutions obtained by Bäcklund transformations.

TABLE 0.1. The Six Painlevé Equations

$$P_{\mathrm{I}}: \quad y'' = 6y^2 + t$$

$$P_{\mathrm{II}}: \quad y'' = 2y^3 + ty + \alpha$$

$$P_{\mathrm{III}}: \quad y'' = \frac{1}{y}(y')^2 - \frac{1}{t}y' + \frac{1}{t}(\alpha y^2 + \beta) + \gamma y^3 + \frac{\delta}{y}$$

$$P_{\mathrm{IV}}: \quad y'' = \frac{1}{2y}(y')^2 + \frac{3}{2}y^3 + 4ty^2 + 2(t^2 - \alpha)y + \frac{\beta}{y}$$

$$P_{\mathrm{V}}: \quad y'' = \left(\frac{1}{2y} + \frac{1}{y-1}\right)(y')^2 - \frac{1}{t}y' + \frac{(y-1)^2}{t^2}\left(\alpha y + \frac{\beta}{y}\right) + \frac{\gamma}{t}y + \delta\frac{y(y+1)}{y-1}$$

$$P_{\mathrm{VI}}: \quad y'' = \frac{1}{2}\left(\frac{1}{y} + \frac{1}{y-1} + \frac{1}{y-t}\right)(y')^2 - \left(\frac{1}{t} + \frac{1}{t-1} + \frac{1}{y-t}\right)y' + \frac{y(y-1)(y-t)}{t^2(t-1)^2}\left(\alpha + \beta\frac{t}{y^2} + \gamma\frac{t-1}{(y-1)^2} + \delta\frac{t(t-1)}{(y-t)^2}\right)$$

In this list, $y = y(t)$ is the dependent variable and $' = d/dt$ stands for the derivative with respect to the independent variable t. The symbols $\alpha, \beta, \ldots$ are parameters.

The first five chapters are the main part of this book. The chapters that follow are an essay on *discrete systems of type A*, which generalize the symmetry structure of P_{II} and P_{IV}. There I give a proof of the Jacobi-Trudi formula in the general setting. (By an "essay" I mean an attempted explanation which some experts might think incomplete.)

I do not think you will find it difficult to read the main part of this book if you know the calculus and linear algebra learnt by first year undergraduates, and have a little knowledge of group theory (such as the definition of a group, and a group defined by generators and fundamental relations). I have chosen the subjects of this book so that you can directly approach the interesting aspects of Painlevé equations without any particular prerequisite. One nice thing about the Painlevé equations is that they are given in explicit forms and one can do computations with them. My hope is that you will try computations by hand (or by computer) while reading this book.

You might expect that a book on Painlevé equations should include such topics as how Painlevé discovered them, what their meaning is and what research has been done so far on them. I did not write much about such things, as they can be found in other books. One example of what I did not explain in this book is:

> Any rational ordinary differential equation of second order that has no movable branching points can be transformed into one of the Painlevé equations, unless it reduces to a simpler equation such as linear equations or differential equations for elliptic functions.

After considering several possibilities, I finally decided it would be very hard to maintain the intended style of this book if I were to include such topics. If you already have some more mathematical background, and want to learn more about Painlevé equations, including their historical aspects, I would recommend this excellent reference:

> Kazuo Okamoto, *Introduction to Painlevé Equations*, Sophia Kokyuroku in Mathematics, No. 19, 1985 (in Japanese)[1].

If you are interested, please browse Section 0 in the appendix (⇒ Appendix 0: Profile of Painlevé equations), where I give some remarks on the origin and the history of Painlevé equations (without detailed explanation of the terminology). In the appendix I have also collected some other topics, including things that are slightly removed from the main flow of this book, that I could not write in sufficient detail in the text, and that I thought to be more advanced. I hope the reader will find the short stories in the appendix helpful.

For several years I have been working with Yasuhiko Yamada, one of my colleagues at Kobe University, on Painlevé equations and related discrete integrable systems. In this book I tried to include extensively what has been made clear through my collaboration with him. The Jacobi-Trudi formula I introduce in Chapter 5, for example, is originally due to Yamada. This book thus contains new material, so in this sense is a *special course* on Painlevé equations rather than a standard textbook. Please keep this in mind as you read this book. I would like to express my thanks to Yasuhiko Yamada, to whom I owe uncountably many valuable suggestions, ever since the planning stages of this book. Without his cooperation, I cannot imagine that this book could have been completed. I am also grateful to the members of the mathematics department at Kobe University, especially to Kyoichi Takano, Kenji Iohara and Jun Nakagawa, who read the manuscript and kindly provided much advice for improving this book.

Masatoshi NOUMI

May, 2000

[1] For English readers, I would suggest: K. Iwasaki, H. Kimura, S. Shimomura and M. Yoshida, *From Gauss to Painlevé — A Modern Theory of Special Functions*, Aspects of Mathematics E16, Vieweg, 1991.

Preface to the English Edition

This book was originally written in Japanese as a volume of the series of mathematical readings "Landscapes of Mathematics" (Asakura Shoten Publishers). My idea in writing this book was to provide an introduction to recent studies on the symmetry of Painlevé equations. I also tried to write the book in a plain language so that it can be read by a reader who has a mathematical background at the undergraduate level.

In order to fulfill this requirement, perhaps I should have added a section on the basic facts concerning groups defined by generators and relations. For those who are not familiar with abstract algebra, the following explanation might be sufficient as a starting point: A *group* is a collection of transformations in which one can compose transformations and take inverse transformations. When one wants to consider several *fundamental transformations* (called the generators) with prescribed *fundamental relations* among them, it is useful to consider the collection of all possible composed transformations among which all the relations are reducible to the fundamental ones. Such a group is the group defined by generators and relations.

As I already mentioned in the preface of the original edition, this book is not meant to be a standard textbook on Painlevé equations, but a special course on their symmetry aspects. It will be my pleasure if this book could help the reader to realize the richness of various mathematical structures appearing there, and in particular the remarkable role of affine Weyl groups in Painlevé equations.

I am grateful to the editors of the AMS for providing me with a chance to publish the English edition of my book. I also express my thanks to Professors Katusmi Nomizu, Wayne Rossman and Nicholas Witte for the many valuable comments they kindly offered during the preparation of the English edition.

Masatoshi NOUMI

September, 2003

CHAPTER 1

What is a Bäcklund Transformation?

The *Bäcklund transformation* is the key for discussing the symmetry of Painlevé equations. In this chapter we take the second Painlevé equation $P_{\rm II}$ as an example for explaining the role of Bäcklund transformations.

1.1. Hamiltonian representation of $P_{\rm II}$

Let us consider the following differential equation for the unknown function $y = y(t)$:

$$P_{\rm II}(b): \qquad y'' = 2y^3 + ty + b - \frac{1}{2}, \tag{1.1}$$

where $' = d/dt$ denotes the derivative with respect to the independent variable t, and $b \in \mathbb{C}$ is a parameter. This differential equation of second order is called the *second Painlevé equation* $P_{\rm II}$. We will also use the notation $P_{\rm II}(b)$ when we need to specify the parameter b.

There are many possible ways to rewrite a differential equation of second order into a system of two differential equations of first order. For $P_{\rm II}$ it is convenient to use the following variables q and p:

$$q = y, \quad p = y' + y^2 + \frac{t}{2}. \tag{1.2}$$

The differential equation $P_{\rm II}(b)$ is then equivalent to the following system of differential equations for q and p:

$$H_{\rm II}(b): \qquad q' = p - q^2 - \frac{t}{2}, \quad p' = 2qp + b. \tag{1.3}$$

In fact, by differentiating the first equation, we have

$$q'' = p' - 2qq' - \frac{1}{2}. \tag{1.4}$$

If we eliminate p and p' in these three equations, we obtain a second order equation for q which recovers $P_{\rm II}(b)$ for $y = q$.

We now define a polynomial $H = H(q, p; t)$ in the three variables (q, p, t) by

$$H = \frac{1}{2}p^2 - \left(q^2 + \frac{t}{2}\right)p - bq = \frac{1}{2}p(p - 2q^2 - t) - bq. \tag{1.5}$$

Since we have

$$\frac{\partial H}{\partial q} = -2qp - b, \quad \frac{\partial H}{\partial p} = p - q^2 - \frac{t}{2}, \quad \frac{\partial H}{\partial t} = -\frac{1}{2}p, \tag{1.6}$$

the differential system $H_{\rm II}(b)$ can be written as

$$q' = \frac{\partial H}{\partial p}, \quad p' = -\frac{\partial H}{\partial q}. \tag{1.7}$$

A system of differential equations of this form is called a *Hamiltonian system* with Hamiltonian H ($\Rightarrow$ Appendix 1: Hamiltonian systems). Summarizing, we have

PROPOSITION 1.1. *The second Painlevé equation P_{II} is equivalent to the Hamiltonian system*

$$H_{\mathrm{II}}: \qquad q' = \frac{\partial H}{\partial p} = p - q^2 - \frac{t}{2}, \quad p' = -\frac{\partial H}{\partial q} = 2qp + b, \tag{1.8}$$

where the Hamiltonian $H = H(q, p; t)$ is given by

$$H = \frac{1}{2}p^2 - \left(q^2 + \frac{t}{2}\right)p - bq = \frac{1}{2}p(p - 2q^2 - t) - bq. \tag{1.9}$$

Let us make a few remarks about the advantages of Hamiltonian systems. Suppose that a solution $q = q(t), p = p(t)$ of the Hamiltonian system (1.8) is given, and define a function

$$h(t) = H(q(t), p(t); t) \tag{1.10}$$

of t, by plugging $q(t)$ and $p(t)$ into the Hamiltonian. For a Hamiltonian system whose Hamiltonian H does not contain the variable t explicitly, H is a first integral, so the function $h(t)$ defined above would be a constant function. This is *not* the case with our H_{II}; we have in fact

$$h'(t) = \frac{\partial H}{\partial t} = -\frac{1}{2}p. \tag{1.11}$$

In the case of a Hamiltonian system whose Hamiltonian depends explicitly on t, the function $h(t)$ can be thought of as a measure indicating how far the system is from a differential system which can be integrated by quadrature. We introduce the *τ-function* $\tau(t)$ for H_{II} by

$$h(t) = \frac{d}{dt}\log\tau(t), \quad \text{i.e.} \quad \tau(t) = \exp\left(\int^t h(t)dt\right). \tag{1.12}$$

Equivalently, we can define the τ-function simply to be a new dependent variable such that

$$H = \frac{d}{dt}\log\tau = \frac{\tau'}{\tau}. \tag{1.13}$$

The τ-function is one of the most important objects in the theory of Painlevé equations. You will come to understand the role of τ-functions gradually as you proceed through this book.

Another advantage of Hamiltonian systems concerns the Poisson structure. For two functions φ and ψ in (q, p, t), define the *Poisson bracket* $\{\varphi, \psi\}$ by

$$\{\varphi, \psi\} = \frac{\partial\varphi}{\partial p}\frac{\partial\psi}{\partial q} - \frac{\partial\varphi}{\partial q}\frac{\partial\psi}{\partial p}. \tag{1.14}$$

Since we have

$$\{\varphi, q\} = \frac{\partial\varphi}{\partial p}, \quad \{\varphi, p\} = -\frac{\partial\varphi}{\partial q} \tag{1.15}$$

in general, the Hamiltonian system (1.8) is expressed as

$$q' = \{H, q\}, \quad p' = \{H, p\} \tag{1.16}$$

in terms of the Poisson bracket. For a general function $\varphi = \varphi(q, p; t)$, the differential equation to be satisfied by the dependent variable $\varphi = \varphi(q(t), p(t); t)$ is then computed as

$$\frac{d\varphi}{dt} = \{H, \varphi\} + \frac{\partial \varphi}{\partial t} \tag{1.17}$$

by the differentiation of compositions of functions. Note that $\frac{\partial}{\partial t}$ in the right-hand side denotes the partial derivative of $\varphi(q, p; t)$ regarded as a function in the three variables (q, p, t). By transformation of variables, the appearance of a differential equation may change into various forms. In terms of the Poisson bracket, however, the differential equation for any function in (q, p, t) can be expressed in the same simple form, which is one of the advantages of a Hamiltonian system.

Suppose now two functions $\widetilde{q} = \widetilde{q}(q, p; t)$ and $\widetilde{p} = \widetilde{p}(q, p; t)$ in (q, p, t) satisfy the relation $\{\widetilde{p}, \widetilde{q}\} = 1$ with respect to the Poisson bracket. Such a pair of functions defines a mapping $(q, p) \to (\widetilde{q}, \widetilde{p})$ whose Jacobian determinant is the constant function 1. If this is the case, the pair $(\widetilde{q}, \widetilde{p})$ is called a *canonical coordinate system* and the mapping $(q, p) \to (\widetilde{q}, \widetilde{p})$ is called a *canonical transformation.* It can be shown that, under a canonical transformation, the differential equation for $(\widetilde{q}, \widetilde{p})$ is again expressed (at least locally) as a Hamiltonian system ($\Rightarrow$ Appendix 2: Poisson structure and canonical transformations).

1.2. Immediately accessible solutions

The second Painlevé equation has some solutions which can be seen immediately from the equation. Let us consider $P_{\mathrm{II}}(b)$ for the case $b = \frac{1}{2}$:

$$P_{\mathrm{II}}\left(\frac{1}{2}\right): \qquad y'' = 2y^3 + ty. \tag{1.18}$$

Noticing that the right-hand side is divisible by y, we clearly see that $y = 0$ (the constant function that is identically zero) is a solution of this equation. Since our equation is nonlinear, it is worthwhile to know it happens to have a solution $y = 0$. In terms of the Hamiltonian system, we have realized that $H_{\mathrm{II}}(\frac{1}{2})$ has a *rational solution* (a solution (q, p) such that both q and p are rational functions in t)

$$q = 0, \quad p = \frac{t}{2} \qquad \left(b = \frac{1}{2}\right). \tag{1.19}$$

We next mention another solution which is slightly more "advanced". Set $b = 0$ in $H_{\mathrm{II}}(b)$, giving

$$H_{\mathrm{II}}(0): \qquad q' = p - q^2 - \frac{t}{2}, \quad p' = 2qp. \tag{1.20}$$

This time the equation for p is divisible by p. Hence it makes sense to search for a solution such that $p = 0$. The resulting equation to be satisfied by q is

$$q' = -q^2 - \frac{t}{2}. \tag{1.21}$$

A differential equation of this form (which requires the derivative of the unknown function to be expressed as a quadratic function of it) is called a *Riccati equation* ($\Rightarrow$ Appendix 3: Riccati equations). Such an equation can be solved by reduction to a linear equation. In the case of (1.21), by the change of variables $q = \frac{u'}{u}$, we obtain a linear equation

$$u'' + \frac{t}{2}u = 0 \tag{1.22}$$

to be satisfied by u. In other words, the Riccati equation (1.21) is *linearized* to (1.22). Equation (1.22), called *Airy's equation*, is one of the basic examples of linear differential equations with irregular singularities. (It has an irregular singular point at $t = \infty$. Probably you need to rescale the variables to get the same Airy equation as you might find in the literature. We will not discuss this particular equation any further. For a description of its solutions we refer the reader to textbooks on ordinary differential equations.) In this way, we see that the Painlevé equation $H_{\text{II}}(0)$ with $b = 0$ has a one-parameter family ($\simeq \mathbb{P}^1$) of solutions

$$q = \frac{u'}{u} = \frac{c_0\varphi_0' + c_1\varphi_1'}{c_0\varphi_0 + c_1\varphi_1}, \quad p = 0 \qquad (b = 0), \tag{1.23}$$

expressed in terms of the general solution $u = c_0\varphi_0 + c_1\varphi_1$ of the linear equation (1.22). Here φ_0 and φ_1 denote two linearly independent solutions of (1.22). Note that every time we specify the ratio of the constants c_0 and c_1 we have precisely one solution q. This means that, on the level of q, we have obtained as many solutions as $\mathbb{P}^1$.

We summarize what we have discussed above about the solutions which can be seen directly from the equation.

PROPOSITION 1.2. *The Hamiltonian representation $H_{\text{II}}(b)$ of the second Painlevé equations has the following particular solutions.*
(1) *When $b = 0$, $H_{\text{II}}(0)$ has a one-parameter family of solutions ($\simeq \mathbb{P}^1$) which are expressed in terms of the solutions of Airy's equation.*
(2) *When $b = \frac{1}{2}$, $H_{\text{II}}(\frac{1}{2})$ has the rational solution $(q, p) = (0, \frac{t}{2})$.*

Please observe which points in the parameter space the solutions (1) and (2) correspond to:

(1.24)

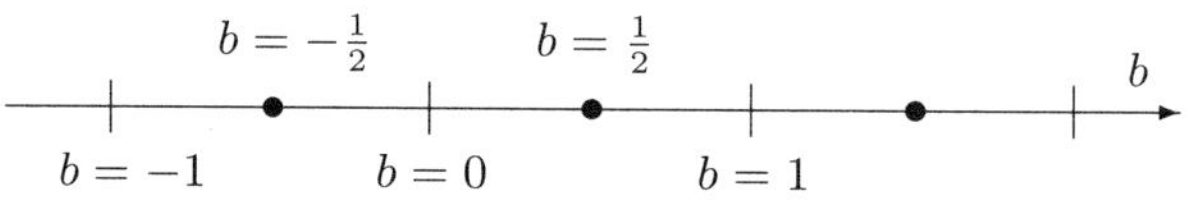

Since the Painlevé equation is of second order, its solutions should be parametrized by two free parameters. Namely, its solution space should form a 2-dimensional manifold, simply a surface. Imagine a picture in which the surface of all solutions of $H_{\text{II}}(b)$ lies over each point b of the parameter space. The surface of solutions over the point $b = 0$ contains $\mathbb{P}^1$ corresponding to the solutions (1), and the surface over the point $b = \frac{1}{2}$ contains a special point corresponding to the rational solution (2).

As we will see below, these are not the only solutions that are either rational or reducible to linear equations. For the Painlevé equation, there is a mechanism by which one can generate new solutions successively from a given solution. Newly generated solutions will then be located at half-integer points or at integer points in the parameter space. Such a mechanism for generating new solutions is the Bäcklund transformation, which we shall discuss in the next section.

1.3. Bäcklund transformations

Let us write again the Hamiltonian representation of the second Painlevé equation:

$$H_{\mathrm{II}}(b): \quad q' = p - q^2 - \frac{t}{2}, \quad p' = 2qp + b. \tag{1.25}$$

Setting

$$\widetilde{q} = q + \frac{b}{p}, \quad \widetilde{p} = p, \tag{1.26}$$

we now try to perform the transformation $(q, p) \longrightarrow (\widetilde{q}, \widetilde{p})$ for this equation. By computing the Poisson bracket, we have $\{\widetilde{p}, \widetilde{q}\} = 1$. Hence we see in advance that this is a canonical transformation. Accordingly we might expect that the resulting equation could again be written as a Hamiltonian system. In fact, we compute

$$\begin{aligned} \widetilde{q}' &= q' - \frac{b}{p^2}p' = p - q^2 - \frac{t}{2} - \frac{b}{p^2}(2q\,p + b) \\ &= p - \left(q + \frac{b}{p}\right)^2 - \frac{t}{2} = \widetilde{p} - \widetilde{q}^2 - \frac{t}{2}, \end{aligned} \tag{1.27}$$

to get the equation for $\widetilde{q}$ of the same form as before. Similarly the equation for $\widetilde{p}$ can be computed as

$$\widetilde{p}' = p' = 2q\,p + b = 2\left(q + \frac{b}{p}\right)p - b = 2\widetilde{q}\,\widetilde{p} - b, \tag{1.28}$$

now having a different sign in front of b. Summarizing, we have seen that the change of variables (1.26) transforms our equation $H_{\mathrm{II}}(b)$ into

$$\widetilde{q}' = \widetilde{p} - \widetilde{q}^2 - \frac{t}{2}, \quad \widetilde{p}' = 2\widetilde{q}\,\widetilde{p} - b. \tag{1.29}$$

(This is again a Hamiltonian system, since the only difference from the original equation is the sign of the parameter.) From these computations we realize:

> For an arbitrary given solution (q, p) of $H_{\mathrm{II}}(b)$, the functions $(\widetilde{q}, \widetilde{p})$ defined by (1.26) give a solution of $H_{\mathrm{II}}(-b)$.

For instance, if we transform the rational solution $(q, p) = (0, \frac{t}{2})$ of $H_{\mathrm{II}}(\frac{1}{2})$, we obtain a rational solution

$$(\widetilde{q}, \widetilde{p}) = \left(\frac{1}{t}, \frac{t}{2}\right) \tag{1.30}$$

of $H_{\mathrm{II}}(-\frac{1}{2})$. This transformation $(q, p) \longrightarrow (\widetilde{q}, \widetilde{p})$ is one of the fundamental Bäcklund transformations for H_{II}. Although we have treated the Hamiltonian system so far, we can easily return to the second order equation for one unknown function. In terms of the original P_{II}, the same transformation gives rise to the change of variables $y \longrightarrow \widetilde{y}$ defined by

$$\widetilde{y} = y + \frac{b}{y' + y^2 + t/2}. \tag{1.31}$$

Notice that this transformation contains the derivative of the unknown function.

For a given differential equation, we call a transformation of the dependent variables a *Bäcklund transformation* of the equation if it leaves the form of the equation invariant.

> **Remark:** The term Bäcklund transformation is often used in a broader sense in the literature. When a differential equation A is transformed into B by a change of the independent and dependent variables, it may be called a Bäcklund transformation from A to B.

A Bäcklund transformation may contain derivatives of the dependent variables. In the cases of equations containing parameters, it may also change the parameters. Recall that, in our example of $H_{\rm II}$, the transformation (1.26) has switched the parameter b to $-b$. Let us now include the change of parameters into the above transformation of variables by setting $\widetilde{b} = -b$, so that

$$\widetilde{q}\,' = \widetilde{p} - \widetilde{q}^2 - \frac{t}{2}, \quad \widetilde{p}\,' = 2\widetilde{q}\,\widetilde{p} + \widetilde{b}. \tag{1.32}$$

The transformation of variables $(q, p, b) \to (\widetilde{q}, \widetilde{p}, \widetilde{b})$ then leaves the equation $H_{\rm II}$ literally invariant. If one regards the parameters as dependent variables whose derivatives are zero, one can also say that a Bäcklund transformation is a transformation of variables for a nonlinear equation under which the equation remains invariant.

There is another fundamental Bäcklund transformation for $H_{\rm II}$. The change of variables

$$\widetilde{q} = -q, \quad \widetilde{p} = -p + 2q^2 + t \tag{1.33}$$

also defines a canonical transformation. The equations satisfied by $\widetilde{q}$ and $\widetilde{p}$ are

$$\widetilde{q}\,' = \widetilde{p} - \widetilde{q}^2 - \frac{t}{2}, \quad \widetilde{p}\,' = 2\widetilde{q}\widetilde{p} + 1 - b, \tag{1.34}$$

by a computation similar to what we did before. If we introduce the change of parameters $\widetilde{b} = 1 - b$, the Hamiltonian system $H_{\rm II}$ remains invariant under the transformation of variables $(q, p, b) \to (\widetilde{q}, \widetilde{p}, \widetilde{b})$.

Since the composition of two Bäcklund transformations is again a Bäcklund transformation, one obtains a number of Bäcklund transformations by repeating the composition successively. This means that various solutions can be generated by Bäcklund transformations from a "seed" solution. We remark that each of the two fundamental Bäcklund transformations for $H_{\rm II}$ discussed above is an *involution*. Namely, the composition with itself goes back to the identity transformation; so, in this case, it is of no use to repeat the same transformation. By combining the two different Bäcklund transformations, however, one can obtain an infinite number of Bäcklund transformations. In the next section, we introduce a notation for simply describing the composition of Bäcklund transformations.

1.4. Composing Bäcklund transformations

Let us consider the first Bäcklund transformation we discussed in the preceding section:

$$(q, p, b) \to (\widetilde{q}, \widetilde{p}, \widetilde{b}): \quad \widetilde{q} = q + \frac{b}{p}, \quad \widetilde{p} = p, \quad \widetilde{b} = -b. \tag{1.35}$$

For a general function $\varphi = \varphi(q, p, b; t)$ of variables (q, p, b, t), we define $s(\varphi)$ by

$$s(\varphi) = \varphi(\widetilde{q}, \widetilde{p}, \widetilde{b}; t) = \varphi\left(q + \frac{b}{p}, p, -b; t\right), \tag{1.36}$$

giving the name s to the operation of replacing every appearance of the variables q, p, b in φ by $\widetilde{q}$, $\widetilde{p}$, $\widetilde{b}$, respectively (with no change in the variable t). We have, in particular, that

$$s(q) = q + \frac{b}{p}, \quad s(p) = p, \quad s(b) = -b, \quad s(t) = t. \tag{1.37}$$

Since s is a change of variables, it preserves the algebraic operations:

$$s(\varphi \pm \psi) = s(\varphi) \pm s(\psi), \quad s(\varphi\psi) = s(\varphi)s(\psi), \quad s\left(\frac{\varphi}{\psi}\right) = \frac{s(\varphi)}{s(\psi)}. \tag{1.38}$$

In the preceding section, we verified that the transformation s leaves the differential system H_{II} invariant; that computation can be expressed as

$$s(q)' = s(p) - s(q)^2 - \frac{t}{2} = s\left(p - q^2 - \frac{t}{2}\right) = s(q') \tag{1.39}$$

with the notation introduced above. Together with the corresponding formulas for p and b, we have

$$s(q)' = s(q'), \quad s(p)' = s(p'), \quad s(b)' = s(b'). \tag{1.40}$$

Hence, for an arbitrary function $\varphi = \varphi(q, p, b; t)$, we have

$$s(\varphi)' = s(\varphi'), \tag{1.41}$$

by the chain rule for differentiating compositions of functions. Namely, the transformation s *commutes* with the operation $'$ of differentiation. We can now redefine a Bäcklund transformation to be *an operation on the variables which commutes with the operation of differentiation.*

Remark: We make a comment for those readers who know the terminology of algebra. Consider the field $K = \mathbb{C}(q, p, b, t)$ of rational functions in the indeterminates q, p, b and t. Then the transformation s can be defined as the automorphism of the field K that maps the generators of K as in (1.37). In view of H_{II}, we define a derivation $'$ on K by setting $q' = p - q^2 + \frac{t}{2}$, $p' = 2qp + b$, $b' = 0$ and $t' = 1$. Endowed with this derivation, K becomes a *differential field*, and s is an automorphism of the differential field K (an automorphism which commutes with the derivation). In this terminology, a Bäcklund transformation for a differential equation can be defined as an automorphism of the differential field that represents the equation.

In the same way let us denote by r the second Bäcklund transformation (1.33) for H_{II}; the rule for replacing the variables is specified by

$$r(q) = -q, \quad r(p) = -p + 2q^2 + t, \quad r(b) = 1 - b, \quad r(t) = t. \tag{1.42}$$

It also commutes with differentiation. We now compute the composition $w = sr$ of the two Bäcklund transformations r and s. Since the composition is again a Bäcklund transformation, for the variables

$$\widetilde{q} = w(q), \quad \widetilde{p} = w(p), \quad \widetilde{b} = w(b), \tag{1.43}$$

we should have the same equation

$$\widetilde{q}' = \widetilde{p} - \widetilde{q}^2 - \frac{t}{2}, \quad \widetilde{p}' = 2\widetilde{q}\widetilde{p} + \widetilde{b}, \tag{1.44}$$

as H_{II}. Since the transformation of b is given by

$$\widetilde{b} = sr(b) = s(1-b) = 1 - s(b) = 1 + b, \tag{1.45}$$

$(\widetilde{q}, \widetilde{p})$ should give rise to a solution of $H_{\mathrm{II}}(b+1)$. The new dependent variables $\widetilde{q}$ and $\widetilde{p}$ are determined explicitly as follows:

$$\begin{aligned} \widetilde{q} &= sr(q) = -s(q) = -q - \frac{b}{p}, \\ \widetilde{p} &= sr(p) = -s(p) + 2s(q)^2 + t = -p + 2\left(q + \frac{b}{p}\right)^2 + t. \end{aligned} \tag{1.46}$$

Let us specialize these variables, for instance, to the rational solution

$$(q, p; b) = \left(0, \frac{t}{2}; \frac{1}{2}\right). \tag{1.47}$$

In this case we have

$$(\widetilde{q}, \widetilde{p}; \widetilde{b}) = \left(-\frac{1}{t}, \frac{t}{2} + \frac{2}{t^2}; \frac{3}{2}\right). \tag{1.48}$$

By plugging these into the equation one can verify that this procedure certainly gives a solution of $H_{\mathrm{II}}(\frac{3}{2})$.

A question arises: How many Bäcklund transformations can we obtain by composing s and r successively? Let us first check the transformation of the parameter b. Since the actions of s and r on b are given by

$$s(b) = -b, \quad r(b) = 1 - b, \tag{1.49}$$

they represent the mirror images (reflections) with respect to $b = 0$ and $b = \frac{1}{2}$, respectively. In fact we have $s^2(b) = b$ and $r^2(b) = b$. Furthermore, we compute

$$s^2(q) = s\left(q + \frac{b}{p}\right) = s(q) + \frac{s(b)}{s(p)} = q + \frac{b}{p} - \frac{b}{p} = q. \tag{1.50}$$

Since we also obviously have $s^2(p) = p$, s^2 is the identity transformation. Together with a similar computation for r^2 we have $s^2 = 1$ and $r^2 = 1$, where "1" stands for the identity transformation. In view of this fact we obtain the following list of all possible compositions of s and r by arranging s and r alternatingly.

$$T_n = (rs)^n, \quad S_n = (rs)^n s \quad (n \in \mathbb{Z}). \tag{1.51}$$

n	…	-2	-1	0	1	2	…
T_n	…	$srsr$	sr	1	rs	$rsrs$	…
S_n	…	$srsrs$	srs	s	r	rsr	…

The actions of T_n and S_n on the parameter can be computed immediately:

$$T_n(b) = b - n, \quad S_n(b) = n - b \qquad (n \in \mathbb{Z}). \tag{1.52}$$

This means that, on the b-line, T_n represents the shift by n, and S_n the reflection with respect to $b = \frac{n}{2}$. The totality of all Bäcklund transformations

$$\widetilde{W} = \langle s, r \rangle = \{\, T_n, S_n \;;\; n \in \mathbb{Z} \,\} \tag{1.53}$$

generated by s and r forms a group with fundamental relations $s^2 = 1$, $r^2 = 1$. This group is called the *affine Weyl group of type* $A_1^{(1)}$. (We use the symbol $\widetilde{W}$, instead of W, for consistency with the notation that will be used later.) In other words, the affine Weyl group acting on the parameter space is lifted to a group of Bäcklund transformations for the second Painlevé equation.

THEOREM 1.3. *The Painlevé equation $H_{\rm II}$ admits the following two fundamental Bäcklund transformations s and r:*

$$(1.54)\qquad \begin{array}{llll} s: & s(q) = q + \dfrac{b}{p}, & s(p) = p, & s(b) = -b, \\ r: & r(q) = -q, & r(p) = -p + 2q^2 + t, & r(b) = 1 - b. \end{array}$$

These two Bäcklund transformations generate a group $\widetilde{W} = \langle s, r\rangle$ with fundamental relations $s^2 = 1$ and $r^2 = 1$.

1.5. Generating solutions

For each Bäcklund transformation $w \in \widetilde{W}$, the variables $\widetilde{q} = w(q)$ and $\widetilde{p} = w(p)$ are rational functions in (q, p, b, t), and $\widetilde{b} = w(b)$ is a linear function in b. If (q, p) is a solution of $H_{\rm II}(b)$, then $(\widetilde{q}, \widetilde{p})$ is a solution of $H_{\rm II}(\widetilde{b})$. An important role of Bäcklund transformations is to generate in this way an infinite number of solutions starting from one solution. The rational solution $(q, p) = (0, \frac{t}{2})$ at $b = \frac{1}{2}$, for instance, gives rise to a rational solution at each half-integer point $b = \frac{1}{2} - n$ $(n \in \mathbb{Z})$ by a Bäcklund transformation[1]. The one-parameter family of Riccati solutions at $b = 0$, in turn, generates a one-parameter family of such solutions at each integer point $b = n$ $(n \in \mathbb{Z})$ that is expressed in terms of Airy functions.

From the viewpoint of special functions it is an interesting problem to find explicit formulas for the rational functions $w(q)$ and $w(p)$. We give below a list of examples to show how q and p are transformed by the T_n and S_n of the preceding section. Also we show some explicit examples of rational solutions obtained by applying them to the rational solution at $b = \frac{1}{2}$.

Table 1.1 is the list of Bäcklund transformations expressed as products of s and r of length ≤ 3. Observe how the expressions rapidly become complex.

Some examples of rational solutions are listed in Table 1.2 in the factored form. Although the list contains only "small" ones, we can already observe some characteristic factors of the rational solutions. Can you see that the polynomials

$$(1.55)\qquad \begin{array}{l} P_1 = 1, \quad P_2 = t, \quad P_3 = t^3 + 4, \quad P_4 = t^6 + 20\,t^3 - 80, \\ P_5 = t^{10} + 60\,t^7 + 11200\,t, \ldots \end{array}$$

in t appear several times, and step by step exchange their parts with each other? By extending the sequence of polynomials by $P_n = P_{1-n}$ $(n \leq 0)$ we see the function p of the rational solution at $b = \frac{1}{2} - n$ can be expressed in the form $P_{n-1}P_{n+1}/2P_n^2$,

[1] It is known that all the rational solutions of $P_{\rm II}$ are obtained in this way. (See Y. Murata: Rational solutions of the second and the fourth equations of Painlevé, Funkcial. Ekvac. **28** (1985), 1–32.)

TABLE 1.1. Bäcklund Transformations for $P_{\rm II}$

	q	p	b
T_{-1}	$-q-\frac{b}{p}$	$-p+2(q+\frac{b}{p})^2+t$	$1+b$
T_0	q	p	b
T_1	$-q+\frac{1-b}{-p+2q^2+t}$	$-p+2q^2+t$	$-1+b$
S_{-1}	$-q-\frac{b}{p}+\frac{1+b}{-p+2(q+\frac{b}{p})^2+t}$	$-p+2(q+\frac{b}{p})^2+t$	$-1-b$
S_0	$q+\frac{b}{p}$	p	$-b$
S_1	$-q$	$-p+2q^2+t$	$1-b$
S_2	$q-\frac{1-b}{-p+2q^2+t}$	$p-2q^2+2(-q+\frac{1-b}{-p+2q^2+t})^2$	$2-b$

TABLE 1.2. Generating Rational Solutions

b	q	p
$\vdots$	$\vdots$	$\vdots$
$-\frac{7}{2}$	$\vdots$	$\frac{(t^3+4)(t^{10}+60t^7+11200t)}{2(t^6+20t^3-80)^2}$
$-\frac{5}{2}$	$\frac{3(t^8+8t^5+160t^2)}{(t^3+4)(t^6+20t^3-80)}$	$\frac{t(t^6+20t^3-80)}{2(t^3+4)^2}$
$-\frac{3}{2}$	$\frac{2(t^3-2)}{t(t^3+4)}$	$\frac{t^3+4}{2t^2}$
$-\frac{1}{2}$	$\frac{1}{t}$	$\frac{t}{2}$
$\frac{1}{2}$	0	$\frac{t}{2}$
$\frac{3}{2}$	$-\frac{1}{t}$	$\frac{t^3+4}{2t^2}$
$\frac{5}{2}$	$-\frac{2(t^3-2)}{t(t^3+4)}$	$\frac{t(t^6+20t^3-80)}{2(t^3+4)^2}$
$\frac{7}{2}$	$-\frac{3(t^8+8t^5+160t^2)}{(t^3+4)(t^6+20t^3-80)}$	$\frac{(t^3+4)(t^{10}+60t^7+11200t)}{2(t^6+20t^3-80)^2}$
$\vdots$	$\vdots$	$\vdots$

for each case in the table. If you wish you can check that p at $b=-\frac{9}{2}$, which is not shown in the table, is also factored as $P_4P_6/2P_5^2$ with

$$P_6 = t^{15} + 140\,t^{12} + 2800\,t^9 + 78400\,t^6 - 3136000\,t^3 - 6272000. \tag{1.56}$$

If you have access to some computer algebra system, you may wish to run your own program to see what is happening.

We now summarize what we have seen in this section.

OBSERVATION 1.4. *There are two sequences of polynomials* P_n *and* Q_n ($n \in \mathbb{Z}$) *in* t *such that the rational solution at* $b = \frac{1}{2} - n$ *is expressed in the form*

$$q = \frac{nQ_n}{P_n P_{n+1}}, \quad p = \frac{P_{n-1}P_{n+1}}{2P_n^2} \tag{1.57}$$

for each $n \in \mathbb{Z}$. *Here* P_n *and* Q_n *are monic polynomials with integer coefficients (with leading coefficient* 1) *and*

$$\begin{aligned} \deg P_n &= \frac{1}{2}n(n-1), \quad P_n = P_{1-n}; \\ \deg Q_n &= n^2 - 1, \quad Q_n = Q_{-n}. \end{aligned} \tag{1.58}$$

We already gave the list of P_n for small n. The polynomials Q_n are given by

$$\begin{aligned} &Q_1 = 1, \quad Q_2 = t^3 - 2, \quad Q_3 = t^8 + 8t^5 + 160t^2, \\ &Q_4 = t^{15} + 50t^{12} + 1000t^9 - 22400t^6 - 112000t^3 - 224000, \dots. \end{aligned} \tag{1.59}$$

The polynomials P_n ($n \in \mathbb{Z}$) are sometimes called the *Yablonski-Vorobiev polynomials*[2]. This remarkable structure of rational solutions that we have seen above is known to be the reflection of a property of the τ-function introduced in Section 1.1. One motivation of this book is to clarify why the Painlevé equations show such a characteristic phenomenon.

Remark: For the purpose of generating special solutions it is more efficient to perform Bäcklund transformations step by step, rather than to apply general formulas for $w(q)$, $w(p)$, as we did in this section. For this point $\Rightarrow$ Appendix 4: Computation of Bäcklund transformations.

1.6. Picture of classical solutions

We now formulate our results about the special solutions of the Painlevé equation P_{II} or H_{II} as a theorem.

THEOREM 1.5. *The Painlevé equation* $H_{\text{II}}(b)$ *has the following special solutions.*
(1) *When the parameter* b *is an integer,* $H_{\text{II}}(n)$ ($n \in \mathbb{Z}$) *has a one-parameter family of solutions expressible rationally in terms of the Airy functions and their derivatives.*
(2) *When* b *is a half-integer,* $H_{\text{II}}(\frac{1}{2} + n)$ ($n \in \mathbb{Z}$) *has a rational solution.*
Furthermore, these solutions are obtained by Bäcklund transformations from those at $b = 0$ *and at* $b = \frac{1}{2}$, *respectively.*

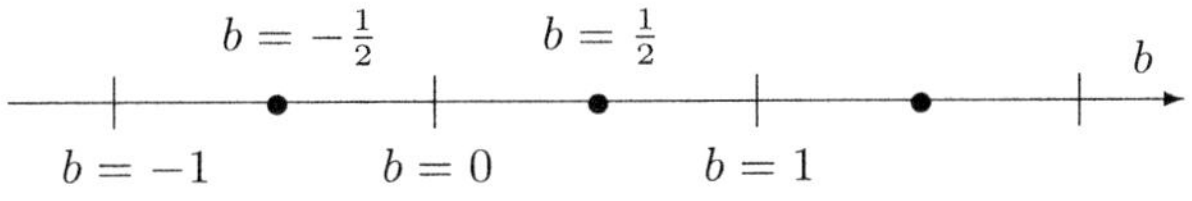

[2] These polynomials P_n are known to be expressed as a specialization of the 2-reduced Schur functions (K. Kajiwara and Y. Ohta: Determinant structure of the rational solutions for the Painlevé II equation, J. Math. Phys. **37** (1996), 4693–4704).

Rational solutions, algebraic solutions and the solutions that are reducible to linear differential equations with rational coefficients are often called the *classical solutions.* The solutions of H_{II} described above are classical solutions in that sense. Ever since the discovery of the Painlevé equations general solutions of each Painlevé equation have been expected to be *far more transcendental* than such classical solutions. It is only since the middle of the 1980s, however, that a modern approach to the transcendence of solutions of the Painlevé equations has been developed. We remark that

(3) Except for those mentioned in (1) and (2) above, all the solutions of H_{II} are *non-classical,*

although we will not go into the details of this subject. There is a tremendous gap between the transcendence of the solutions in (1) and (2), and that of all the others. (⇒ Appendix 5: Classical solutions and invariant divisors.)

Recall that each of the Bäcklund transformations S_n ($n \in \mathbb{Z}$), introduced in Section 1.4, acts on the parameter space as the reflection with respect to the point $b = \frac{n}{2}$. We remark that the fixed points of S_n ($n \in \mathbb{Z}$) are exactly the places where the classical solutions of H_{II} arise. To be more precise, the integer points $b = n$ and the half-integer points $b = \frac{1}{2} + n$ should be distinguished since the two types of classical solutions have different natures; we will come back to this point later.

CHAPTER 2

The Symmetric Form

The main object of this chapter is the fourth Painlevé equation $P_{\rm IV}$, which now contains two parameters α and β. This equation can be represented as a Hamiltonian system just as in the case of $P_{\rm II}$, and a story about $P_{\rm IV}$ can be developed similarly to that of $P_{\rm II}$. In this chapter, however, we approach $P_{\rm IV}$ from a slightly different point of view.

2.1. The symmetric form of $P_{\rm IV}$

Let us consider the following system of ordinary differential equations for three dependent variables f_0, f_1 and f_2:

$$\begin{cases} f_0' = f_0(f_1 - f_2) + \alpha_0, \\ f_1' = f_1(f_2 - f_0) + \alpha_1, \\ f_2' = f_2(f_0 - f_1) + \alpha_2, \end{cases} \tag{2.1}$$

where $' = d/dt$ is the derivative with respect to the independent variable t, and α_0, α_1 and α_2 are parameters. Note first that (2.1) has homogeneity; it is invariant under the rescaling of variables $t \to t/c$, $f_j \to cf_j$ and $\alpha_j \to c^2\alpha_j$ $(j = 0, 1, 2)$.

This differential system of third order has a simple integral. In fact, by taking the sum of the three equations, we have

$$(f_0 + f_1 + f_2)' = \alpha_0 + \alpha_1 + \alpha_2 = k \quad \text{(constant)}; \tag{2.2}$$

hence

$$f_0 + f_1 + f_2 = k\,t + c, \tag{2.3}$$

with an integration constant c. We assume the generic situation $k \neq 0$, leaving aside for now the case $k = 0$. Then, in view of the freedom in rescaling the variables as above, we have only to consider the case $k = 1$. Let us normalize so that

$$\alpha_0 + \alpha_1 + \alpha_2 = 1, \qquad f_0 + f_1 + f_2 = t, \tag{2.4}$$

by eliminating also the freedom of parallel translation with respect to t. Under this normalization, (2.1) turns out to be equivalent to the fourth Painlevé equation

$$P_{\rm IV}: \quad y'' = \frac{1}{2y}(y')^2 + \frac{3}{2}y^3 + 4\,t\,y^2 + 2(t^2 - \alpha)y + \frac{\beta}{y}. \tag{2.5}$$

PROPOSITION 2.1. *Under the normalization condition (2.4), the system of differential equations (2.1) is equivalent to $P_{\rm IV}$.*

Since $f_0 = t - f_1 - f_2$ by (2.4), we first eliminate f_0 to get

$$\begin{cases} f_1' = f_1(f_1 + 2f_2 - t) + \alpha_1, \\ f_2' = f_2(t - 2f_1 - f_2) + \alpha_2. \end{cases} \tag{2.6}$$

From these two equations, together with the one obtained by differentiating the equation for f_1, we can eliminate f_2 and f_2'. The resulting equation for $y = f_1$ is

$$y'' = \frac{1}{2y}(y')^2 + \frac{3}{2}y^3 - 2\,t\,y^2 + \left(\frac{t^2}{2} - \alpha_0 + \alpha_2\right) y - \frac{\alpha_1^2}{2y}. \tag{2.7}$$

Some coefficients in this equation are slightly different from those in P_{IV} quoted above. We can, however, apply the change of variables $t \to \sqrt{2}t$, $y \to -y/\sqrt{2}$ to (2.7) to get

$$y'' = \frac{1}{2y}(y')^2 + \frac{3}{2}y^3 + 4\,t\,y^2 + 2(t^2 - \alpha_0 + \alpha_2)y - \frac{2\alpha_1^2}{y}, \tag{2.8}$$

which is precisely P_{IV} with parameters $\alpha = \alpha_0 - \alpha_2$, $\beta = -2\alpha_1^2$. In what follows we call the differential system (2.1) with normalization (2.4) the *symmetric form* of the fourth Painlevé equation[1].

Note that our symmetric form has a manifest symmetry with respect to rotating the indices $0, 1, 2$ in this order. It means that the transformation π of variables and parameters defined by

$$\begin{aligned} &\pi(f_0) = f_1, \quad \pi(f_1) = f_2, \quad \pi(f_2) = f_0, \\ &\pi(\alpha_0) = \alpha_1, \quad \pi(\alpha_1) = \alpha_2, \quad \pi(\alpha_2) = \alpha_0 \end{aligned} \tag{2.9}$$

gives a Bäcklund transformation for the symmetric form of P_{IV}.

We remark that when $\alpha_0 = \alpha_1 = \alpha_2 = 0$ the differential system (2.1) reduces to the *Lotka-Volterra equation*

$$f_0' = f_0(f_1 - f_2), \quad f_1' = f_1(f_2 - f_0), \quad f_2' = f_2(f_0 - f_1), \tag{2.10}$$

which is a competition model for three species. In other words, the Painlevé equation P_{IV} can be regarded as a variation of the Lotka-Volterra equation obtained by introducing parameters.

2.2. Fundamental special solutions

The symmetric form of P_{IV} has several special solutions which can easily be recognized. One solution arises from the case where everything is symmetric with respect to the three indices. If we impose the condition $\alpha_0 = \alpha_1 = \alpha_2$ and $f_0 = f_1 = f_2$, such a solution is determined to be

$$(\alpha_0, \alpha_1, \alpha_2; f_0, f_1, f_2) = \left(\frac{1}{3}, \frac{1}{3}, \frac{1}{3}; \frac{t}{3}, \frac{t}{3}, \frac{t}{3}\right). \tag{2.11}$$

Compared with the case of P_{II}, this can be thought of as corresponding to the rational solution $(q, p) = (0, \frac{t}{2})$ at $b = \frac{1}{2}$.

[1] The description of symmetry of P_{IV} in terms of the symmetric form, as in this chapter, is due to M. Noumi and Y. Yamada [Symmetries in the fourth Painlevé equation and Okamoto polynomials, Nagoya Math. J. **153** (1999), 53–86]. The fact that P_{IV} is equivalently expressed as (2.1) had been recognized earlier by V.E. Adler [Nonlinear chains and Painlevé equation, Physica D **73** (1994), 335–351].

The other solutions to be discussed in this section are of Riccati type. Consider the first equation

$$f_0' = f_0(f_1 - f_2) + \alpha_0. \tag{2.12}$$

When $\alpha_0 = 0$ the right-hand side is divisible by f_0. Hence it makes sense to specialize by setting $f_0 = 0$. Since $f_1 + f_2 = t$ in this case, we obtain

$$f_1' = f_1 f_2 + \alpha_1 = f_1(t - f_1) + \alpha_1, \tag{2.13}$$

which is a Riccati equation for f_1. By the transformation of variables $f_1 = u'/u$, we obtain the linear differential equation for u

$$u'' - t\,u' - \alpha_1 u = 0, \tag{2.14}$$

which is called *Hermite's differential equation.* (Use the transformation $t \to \sqrt{2}\,t$ to get the same Hermite equation as in the literature.) In this way we see that (2.1) has a one-parameter family of solutions

$$f_0 = 0, \quad f_1 = \frac{\varphi'}{\varphi}, \quad f_2 = t - \frac{\varphi'}{\varphi} \tag{2.15}$$

expressed in terms of the general solutions φ of Hermite's equation (2.14). We remark that although the solutions of this class are generally transcendental, rational solutions are included as well when α_1 takes integer values. In fact (2.14) has polynomial solutions (Hermite polynomials) for $\alpha_1 = -n$ $(n = 0, 1, 2, \ldots)$; if φ is a polynomial, both f_1 and f_2 are rational functions. Similarly, when either $\alpha_1 = 0$ or $\alpha_2 = 0$, the symmetric form (2.1) of P_{IV} has a one-parameter family of Riccati type solutions obtained by the specialization $f_1 = 0$ or by $f_2 = 0$, respectively. Compared with the case of P_{II}, these solutions correspond to the solutions expressed by Airy functions at $b = 0$. When $\alpha_0 = 0$ *and* $\alpha_1 = 0$ the two families of Riccati type solutions, one coming from $\alpha_0 = 0$ and the other from $\alpha_1 = 0$, have one rational solution

$$(\alpha_0, \alpha_1, \alpha_2; f_0, f_1, f_2) = (0, 0, 1; 0, 0, t) \tag{2.16}$$

in common.

The parameter space for P_{IV} is a plane (2-dimensional affine space), while that for P_{II} is a line. Let us try to visualize the parameter space to see where the special solutions mentioned above are located. Our parameters are specified by the triple of functions $(\alpha_0, \alpha_1, \alpha_2)$ satisfying the linear equation $\alpha_0 + \alpha_1 + \alpha_2 = 1$. If we regard α_j $(j = 0, 1, 2)$ as real-valued functions, then $(\alpha_0, \alpha_1, \alpha_2)$ can be interpreted as a *triangular coordinate system* of the plane. Fix a regular triangle $\Delta P_0 P_1 P_2$ of height 1 on the plane, as in Figure 2.1. For a given point P on the plane we measure the distance between the line $P_1 P_2$ and P, taking the direction towards P_0, and call it α_0. In other words, α_0 is the linear function that takes the value 0 on the line $P_1 P_2$ and the value 1 at P_0. The functions α_1, α_2 are defined similarly by measuring the distances from $P_2 P_0$ and from $P_0 P_1$ respectively. Then we have $\alpha_0 + \alpha_1 + \alpha_2 = 1$, as follows from elementary geometry. Conversely, for an arbitrary triple $(\alpha_0, \alpha_1, \alpha_2)$ of real numbers satisfying this condition, one can find a unique point P on the plane corresponding to the triple. For example the triangular coordinates for P_0, P_1, P_2 are given by $(1, 0, 0)$, $(0, 1, 0)$, $(0, 0, 1)$. The barycenter of the triangle $\Delta P_0 P_1 P_2$ has the coordinates $(\frac{1}{3}, \frac{1}{3}, \frac{1}{3})$.

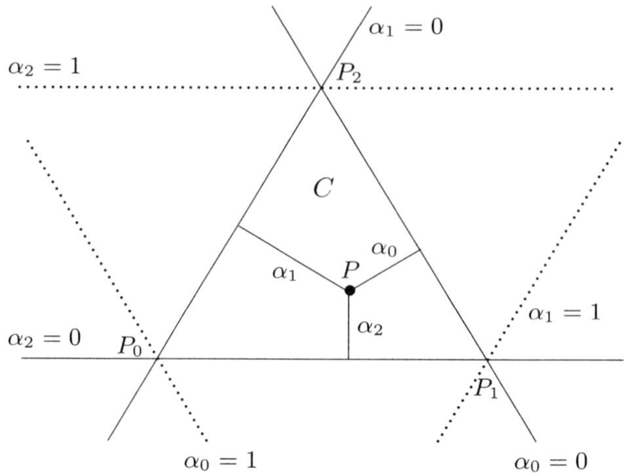

FIGURE 2.1. Triangular Coordinate System

On the parameter space, visualized as a plane with triangular coordinates, the rational solution

$$(\alpha_0, \alpha_1, \alpha_2; f_0, f_1, f_2) = \left(\frac{1}{3}, \frac{1}{3}, \frac{1}{3}; \frac{t}{3}, \frac{t}{3}, \frac{t}{3}\right) \tag{2.17}$$

of the symmetric form of P_{IV} is located at the barycenter of $\Delta P_0P_1P_2$. On the other hand, along each of the three lines $\alpha_0 = 0$, $\alpha_1 = 0$, $\alpha_2 = 0$ bounding the triangle $\Delta P_0P_1P_2$, there is a one-parameter family of Riccati type solutions. Recall that for each $j = 0, 1, 2$ the Riccati type solutions along $\alpha_j = 0$ are obtained by the specialization $f_j = 0$. In this sense the three dependent variables f_0, f_1, f_2 of the symmetric form are related to the lines $\alpha_0 = 0$, $\alpha_1 = 0$, $\alpha_2 = 0$ respectively. (According to the terminology of H. Umemura, f_j is the *invariant divisor* along $\alpha_j = 0$. $\Rightarrow$ Appendix 5: Classical solutions and invariant divisors.) Also at each of the three vertices P_0, P_1 or P_2 two one-parameter families of Riccati type solutions intersect, and give rise to a rational solution

$$(\alpha_0, \alpha_1, \alpha_2; f_0, f_1, f_2) = \begin{cases} (1, 0, 0; t, 0, 0), \\ (0, 1, 0; 0, t, 0), \quad \text{or} \\ (0, 0, 1; 0, 0, t), \end{cases} \tag{2.18}$$

respectively.

The Bäcklund transformation π we have seen in the preceding section corresponds to the rotation by 120° around the barycenter of $\Delta P_0P_1P_2$. If we understand the whole picture of Bäcklund transformations for our symmetric form, we will be able to use these transformations to produce a large class of classical solutions from the basic special solutions discussed in this section. Before doing this, let us first investigate the structure of the affine Weyl group acting on this parameter space.

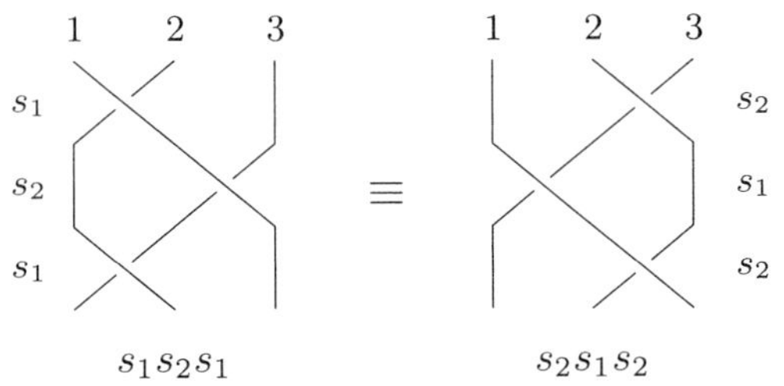

FIGURE 2.2. A Braid Relation

2.3. Affine Weyl group

Underlying the symmetry of the fourth Painlevé equation is the affine Weyl group of type $A_2^{(1)}$. We first define this group in terms of generators and fundamental relations. The *affine Weyl group* of type $A_2^{(1)}$ is the group $W = \langle s_0, s_1, s_2 \rangle$ generated by three elements s_0, s_1, s_2 with fundamental relations

$$s_j^2 = 1, \quad (s_j s_{j+1})^3 = 1 \quad (j = 0, 1, 2). \tag{2.19}$$

Hereafter we use the convention $s_{j+3} = s_j$ ($j \in \mathbb{Z}$) regarding the indices j for s_j as elements of $\mathbb{Z}/3\mathbb{Z}$. A relation of the form $(s_1s_2)^3 = 1$, which specifies the order of products of two elements, is called a *Coxeter relation.* Under the condition $s_1^2 = s_2^2 = 1$, this relation can also be written as $s_1s_2s_1 = s_2s_1s_2$; the latter is called a *braid relation* (Figure 2.2).

The group of Bäcklund transformations for the fourth Painlevé equation is actually a bigger group than W. It is the *extended affine Weyl group* $\widetilde{W} = \langle s_0, s_1, s_2; \pi \rangle$ obtained by adjoining a new element π with relations

$$\pi^3 = 1, \quad \pi s_j = s_{j+1}\pi \quad (j = 0, 1, 2). \tag{2.20}$$

In this section we offer a standard explanation to realize this group geometrically.

> **Remark:** From this definition, one has $\widetilde{W} \simeq W \rtimes \Omega$, $\Omega = \{1, \pi, \pi^2\} \simeq \mathbb{Z}/3\mathbb{Z}$. Namely, π acts on W as the automorphism of order 3 that rotates the indices of s_j; $\widetilde{W}$ is the semidirect product of W and the cyclic group Ω of order 3 acting on W. (⇒ Appendix 6: Semidirect product of groups.)

In terms of the triangular coordinates we discussed in the preceding section, let us consider the three families of lines $\alpha_0 = n$, $\alpha_1 = n$ and $\alpha_2 = n$ ($n \in \mathbb{Z}$). In this way, we obtain a triangular lattice in the plane, as in Figure 2.3. We call s_0, s_1 and s_2 the reflections (symmetry transformation) with respect to the lines $\alpha_0 = 0$, $\alpha_1 = 0$ and $\alpha_2 = 0$, respectively. Then we denote by

$$W = \langle s_0, s_1, s_2 \rangle \tag{2.21}$$

the group of all transformations on the plane obtained by composition from s_0, s_1, s_2. This W gives in fact a realization of the affine Weyl group of type $A_2^{(1)}$. Since each of s_0, s_1, s_2 is a symmetry transformation with respect to a line, the composition of s_i with itself gives the identity transformation:

$$s_0^2 = 1, \quad s_1^2 = 1, \quad s_2^2 = 1. \tag{2.22}$$

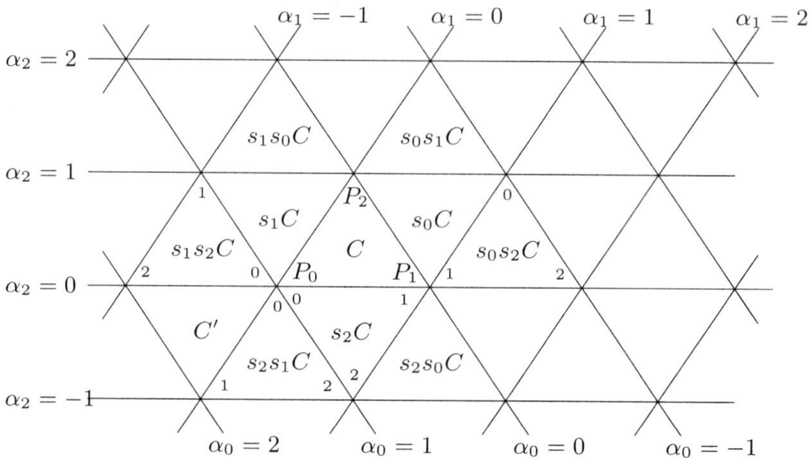

FIGURE 2.3. The Parameter Space of P_{IV}

Also, the composition s_1s_2 of s_2 and s_1 (in that order) represents the rotation by 120° around P_0, as can be seen from the figure. Hence we have $(s_1s_2)^3 = 1$. Similarly we have

$$(s_1s_2)^3 = 1, \quad (s_2s_0)^3 = 1, \quad (s_0s_1)^3 = 1. \tag{2.23}$$

It is known that (2.22) and (2.23) give the whole set of fundamental relations for the transformation group generated by the three reflections s_0, s_1, s_2; we will not enter the proof of this fact. In Figure 2.3 we denote the regular triangle $\triangle P_0P_1P_2$ by C and show how it is mapped by s_j and s_is_j. If we apply all the transformations $w \in W$ to make triangles $w\,C$, they cover the whole plane. Furthermore this procedure gives a one-to-one correspondence between the elements $w \in W$ and the triangular tiles that cover the plane. For instance, the triangle C' in the figure can be expressed in two ways as $s_1s_2s_1C = s_2s_1s_2C$; two expressions $s_1s_2s_1$, $s_2s_1s_2$ represent the same element in W, as is seen from the fundamental relations. Imagine you are looking into a kaleidoscope with three mirrors.

We introduce another transformation π, the rotation by 120° anticlockwise around the barycenter of C. By adjoining π to W, we set

$$\widetilde{W} = \langle s_0, s_1, s_2, \pi \rangle. \tag{2.24}$$

Then it can be seen graphically that the following relations hold:

$$\pi^3 = 1, \quad \pi s_j = s_{j+1}\pi \quad (j = 0, 1, 2). \tag{2.25}$$

It is known that the relations (2.22), (2.23), (2.25) give the whole set of fundamental relations for the group generated by the transformations s_0, s_1, s_2, π; they are precisely the fundamental relations for our extended affine Weyl group.

We now reformulate the realization of $\widetilde{W}$ of this section in terms of the triangular coordinates, which are the parameters for the symmetric form of the Painlevé equation. We identify a point P on the plane with the vector $\boldsymbol{a} = (a_0, a_1, a_2)$ representing the triangular coordinates of P. Then the transformations s_0, s_1, s_2,

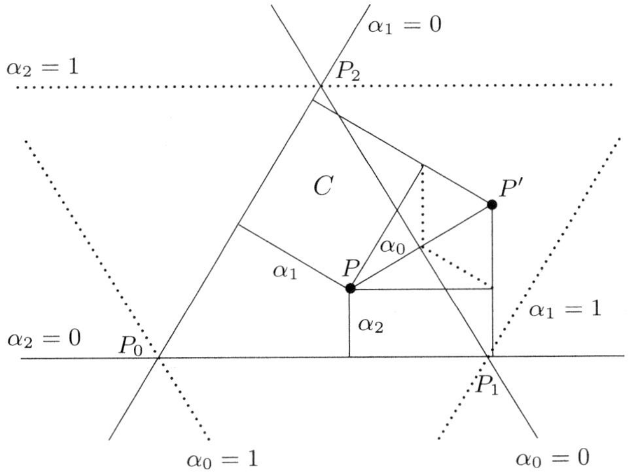

FIGURE 2.4. The Reflection s_0

π induce the following transformations of the triangular coordinates, respectively (see Figure 2.4):

$$\begin{aligned} s_0.\boldsymbol{a} &= (-a_0, a_1 + a_0, a_2 + a_0), \\ s_1.\boldsymbol{a} &= (a_0 + a_1, -a_1, a_2 + a_1), \\ s_2.\boldsymbol{a} &= (a_0 + a_2, a_1 + a_2, -a_2), \\ \pi.\boldsymbol{a} &= (a_2, a_0, a_1). \end{aligned} \tag{2.26}$$

In what follows, we complexify the situation and consider the affine subspace

$$V = \{\, \boldsymbol{a} = (a_0, a_1, a_2) \in \mathbb{C}^3 \,;\, a_0 + a_1 + a_2 = 1 \,\} \tag{2.27}$$

of the 3-dimensional complex affine space $\mathbb{C}^3$. In this setting the parameters α_0, α_1, α_2 for the symmetric form of P_{IV} are functions on V that take the values

$$\alpha_0(\boldsymbol{a}) = a_0, \quad \alpha_1(\boldsymbol{a}) = a_1, \quad \alpha_2(\boldsymbol{a}) = a_2 \tag{2.28}$$

at each point $\boldsymbol{a} = (a_0, a_1, a_2)$, respectively; accordingly, they satisfy the linear relation $\alpha_0 + \alpha_1 + \alpha_2 = 1$ on V. We define the transformations s_0, s_1, s_2 and π of V by (2.26), and set $\widetilde{W} = \langle s_0, s_1, s_2, \pi \rangle$, so that $\widetilde{W}$ becomes a group of affine transformations of V. Supposing that a function β on V and an element $w \in \widetilde{W}$ are given, we define the action $w(\beta)$ of $w \in \widetilde{W}$ on β to be the function such that

$$w(\beta)(\boldsymbol{a}) = \beta(w^{-1}.\boldsymbol{a}) \qquad (\boldsymbol{a} \in V). \tag{2.29}$$

(We have used w^{-1} in the right-hand side so that the composition of transformations preserves the order of multiplication as in $\widetilde{W}$. ⇒ Appendix 4: Computation of Bäcklund transformations.) From this definition the action of each generator of $\widetilde{W}$

on α_j is computed as follows:

(2.30)

	α_0	α_1	α_2
s_0	$-\alpha_0$	$\alpha_1+\alpha_0$	$\alpha_2+\alpha_0$
s_1	$\alpha_0+\alpha_1$	$-\alpha_1$	$\alpha_2+\alpha_1$
s_2	$\alpha_0+\alpha_2$	$\alpha_1+\alpha_2$	$-\alpha_2$
π	α_1	α_2	α_0

In general, suppose that a function φ on V is expressed as $\varphi=\varphi(\alpha_0,\alpha_1,\alpha_2)$ in terms of the triangular coordinates. Then the action of $w\in\widetilde{W}$ on φ is given by

$$w(\varphi)=\varphi(w(\alpha_0),w(\alpha_1),w(\alpha_2)). \tag{2.31}$$

Hence $w(\varphi)$ can be determined completely by the table (2.30).

Remark: Let us give some examples of how $s_i(\alpha_j)$, $\pi(\alpha_j)$ are computed. For $a=(a_0,a_1,a_2)$ given we have $s_0^{-1}.a=s_0.a=(-a_0,a_1+a_0,a_2+a_0)$; hence

$$s_0(\alpha_1)(a)=\alpha_1(s_0^{-1}.a)=a_1+a_0=\alpha_1(a)+\alpha_0(a). \tag{2.32}$$

This means that $s_0(\alpha_1)=\alpha_1+\alpha_0$. From $\pi^{-1}.a=(a_1,a_2,a_0)$ we obtain

$$\pi(\alpha_1)(a)=\alpha_1(\pi^{-1}.a)=a_2=\alpha_2(a); \tag{2.33}$$

hence $\pi(\alpha_1)=\alpha_2$.

THEOREM 2.2. *Define the transformations s_0,s_1,s_2,π of the triangular coordinates $\alpha_0,\alpha_1,\alpha_2$ $(\alpha_0+\alpha_1+\alpha_2=1)$ of the plane as in the table* (2.30). *Then the fundamental relations among s_0,s_1,s_2,π are given by*

$$s_j^2=1,\quad (s_js_{j+1})^3=1;\quad \pi^3=1,\quad \pi s_j=s_{j+1}\pi\quad (j=0,1,2). \tag{2.34}$$

Namely, the group $\widetilde{W}=\langle\, s_0,s_1,s_2,\pi\,\rangle$ generated by these transformations provides a realization of the extended affine Weyl group of type $A_2^{(1)}$.

At this point we make some remarks concerning the *Cartan matrix* and the *Dynkin diagram*, with which one can neatly describe the action (2.30) of s_i on α_j. Observe that, for any i,j, the difference of $s_i(\alpha_j)$ and α_j is an integer multiple of α_i. Writing $s_i(\alpha_j)$ in the form

$$s_i(\alpha_j)=\alpha_j-\alpha_i a_{ij}\qquad (i,j=0,1,2), \tag{2.35}$$

we obtain a 3×3 symmetric matrix $A=(a_{ij})_{i,j=0}^2$:

$$A=\begin{bmatrix} 2 & -1 & -1\\ -1 & 2 & -1\\ -1 & -1 & 2\end{bmatrix}. \tag{2.36}$$

This matrix A is called the *Cartan matrix* of type $A_2^{(1)}$; if you remember this matrix A, you can immediately reproduce the tables (2.30). The Dynkin diagram provides a way to visualize the matrix A. To each index $0,1,2$, we assign a node (circle) $\circ$, and for each pair (i,j) with $a_{ij}=a_{ji}=-1$ we join the two nodes that correspond to i and j by an edge as o—o. For the Cartan matrix A above we obtain the following figure, called the *Dynkin diagram* of type $A_2^{(1)}$:

(2.37)

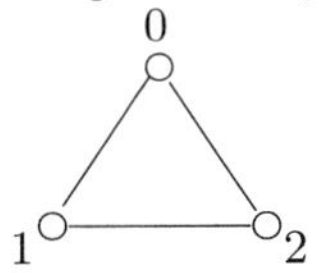

Since the diagonal entries of A are all 2, one can recover A from this diagram. (For general cases ⇒ Appendix 7: Cartan matrices and Dynkin diagrams.)

2.4. Bäcklund transformations

Let us recall the symmetric form of the fourth Painlevé equation:

$$f_j' = f_j(f_{j+1} - f_{j+2}) + \alpha_j \quad (j = 0, 1, 2); \quad f_0 + f_1 + f_2 = t. \tag{2.38}$$

Hereafter we regard the indices j for f_j as elements of $\mathbb{Z}/3\mathbb{Z}$ so that $f_{j+3} = f_j$. Modifying the dependent variables f_0, f_1, f_2, we define new dependent variables g_0, g_1, g_2 by setting

$$g_0 = f_0, \quad g_1 = f_1 + \frac{\alpha_0}{f_0}, \quad g_2 = f_2 - \frac{\alpha_0}{f_0}, \tag{2.39}$$

and determine the system of differential equations to be satisfied by g_0, g_1, g_2. Since the inverse transformation is given by

$$f_0 = g_0, \quad f_1 = g_1 - \frac{\alpha_0}{g_0}, \quad f_2 = g_2 + \frac{\alpha_0}{g_0}, \tag{2.40}$$

we compute as follows:

$$\begin{aligned} g_0' = f_0' &= f_0(f_1 - f_2) + \alpha_0 \\ &= g_0\left(g_1 - \frac{\alpha_0}{g_0} - g_2 - \frac{\alpha_0}{g_0}\right) + \alpha_0 \\ &= g_0(g_1 - g_2) - \alpha_0. \end{aligned} \tag{2.41}$$

Similarly, we have

$$\begin{aligned} g_1' &= \left(f_1 + \frac{\alpha_0}{f_0}\right)' = f_1' - \frac{\alpha_0}{f_0^2} f_0' \\ &= f_1(f_2 - f_0) + \alpha_1 - \frac{\alpha_0}{f_0}(f_1 - f_2) - \frac{\alpha_0^2}{f_0^2} \\ &= \left(g_1 - \frac{\alpha_0}{g_0}\right)\left(g_2 - g_0 + \frac{\alpha_0}{g_0}\right) + \alpha_1 - \frac{\alpha_0}{g_0}\left(g_1 - g_2 - 2\frac{\alpha_0}{g_0}\right) - \frac{\alpha_0^2}{g_0^2} \\ &= g_1(g_2 - g_0) + \alpha_1 + \alpha_0. \end{aligned} \tag{2.42}$$

Together with the computation for g_2' we finally obtain

$$\begin{cases} g_0' = g_0(g_1 - g_2) - \alpha_0, \\ g_1' = g_1(g_2 - g_0) + \alpha_1 + \alpha_0, \\ g_2' = g_2(g_0 - g_1) + \alpha_2 + \alpha_0. \end{cases} \tag{2.43}$$

This means that if we define the parameters $\beta_0, \beta_1, \beta_2$ by

$$\beta_0 = -\alpha_0, \quad \beta_1 = \alpha_1 + \alpha_0, \quad \beta_2 = \alpha_2 + \alpha_0, \tag{2.44}$$

TABLE 2.1. Bäcklund Transformations for the Symmetric Form of P_{IV}

	α_0	α_1	α_2	f_0	f_1	f_2
s_0	$-\alpha_0$	$\alpha_1+\alpha_0$	$\alpha_2+\alpha_0$	f_0	$f_1+\dfrac{\alpha_0}{f_0}$	$f_2-\dfrac{\alpha_0}{f_0}$
s_1	$\alpha_0+\alpha_1$	$-\alpha_1$	$\alpha_2+\alpha_1$	$f_0-\dfrac{\alpha_1}{f_1}$	f_1	$f_2+\dfrac{\alpha_1}{f_1}$
s_2	$\alpha_0+\alpha_2$	$\alpha_1+\alpha_2$	$-\alpha_2$	$f_0+\dfrac{\alpha_2}{f_2}$	$f_1-\dfrac{\alpha_2}{f_2}$	f_2
π	α_1	α_2	α_0	f_1	f_2	f_0

then g_0, g_1, g_2 satisfy the system of equations

$$g_j' = g_j(g_{j+1} - g_{j+2}) + \beta_j \qquad (j = 0, 1, 2); \quad g_0 + g_1 + g_2 = t \tag{2.45}$$

of the same form as in (2.38). In this sense, the transformation of variables $f_j \to g_j$, $\alpha_j \to \beta_j$ $(j = 0, 1, 2)$ provides a Bäcklund transformation for the symmetric form.

We note that the transformation of the parameters (2.44) is precisely the action of s_0 we have seen in the previous section: $\beta_j = s_0(\alpha_j)$ $(j = 0, 1, 2)$. In view of this we refer to the Bäcklund transformation described above as s_0:

$$\begin{aligned} &s_0(\alpha_0) = -\alpha_0, && s_0(\alpha_1) = \alpha_1 + \alpha_0, && s_0(\alpha_2) = \alpha_2 + \alpha_0, \\ &s_0(f_0) = f_0, && s_0(f_1) = f_1 + \frac{\alpha_0}{f_0}, && s_0(f_2) = f_2 - \frac{\alpha_0}{f_0}. \end{aligned} \tag{2.46}$$

(When we work with transformations of variables, we will use the notation $s_0(\varphi)$ rather than $s_0.\varphi$ although the two have the same meaning.) Since the symmetric form has the obvious symmetry with respect to the rotation of indices, we can infer the two other Bäcklund transformations s_1, s_2, too. By means of the Bäcklund transformation

$$\pi(\alpha_j) = \alpha_{j+1}, \quad \pi(f_j) = f_{j+1} \qquad (j = 0, 1, 2), \tag{2.47}$$

we define s_1 and s_2 by

$$s_1 = \pi s_0 \pi^{-1}, \quad s_2 = \pi s_1 \pi^{-1}. \tag{2.48}$$

Taking all of these together, we finally obtain Bäcklund transformations as listed in Table 2.1. We remark that the action of s_i on f_j is expressed in the form

$$s_i(f_j) = f_j + \frac{\alpha_i}{f_i} u_{ij} \qquad (i, j = 0, 1, 2). \tag{2.49}$$

If we set $U = (u_{ij})_{i,j=0}^2$, then U is the following skew-symmetric matrix:

$$U = \begin{bmatrix} 0 & 1 & -1 \\ -1 & 0 & 1 \\ 1 & -1 & 0 \end{bmatrix}. \tag{2.50}$$

THEOREM 2.3. *The transformations s_0, s_1, s_2, π defined as in Table 2.1 are Bäcklund transformations for the symmetric form of the fourth Painlevé equation $P_{\rm IV}$. Furthermore, these transformations satisfy the following relations*:

$$s_j^2 = 1, \quad (s_j s_{j+1})^3 = 1; \quad \pi^3 = 1, \quad \pi s_j = s_{j+1}\pi \quad (j = 0, 1, 2). \tag{2.51}$$

This theorem means that the extended affine Weyl group of type $A_2^{(1)}$ acts on the symmetric form of $P_{\rm IV}$ as a group of Bäcklund transformations $\widetilde{W} = \langle s_0, s_1, s_2, \pi \rangle$.

In order to verify the fundamental relations of the extended affine Weyl group, we have only to compute honestly according to Table 2.1. Since we have already completed the task with respect to the action on α_j in the previous section, we need only to consider the action on f_j. Also, thanks to the rotation symmetry it suffices to verify $s_0^2 = 1$ and $(s_0 s_1)^3 = 1$. First, let us check $s_0^2 = 1$. For $j = 0, 1, 2$ we have

$$s_0^2(f_j) = s_0\left(f_j + \frac{\alpha_0}{f_0}u_{0j}\right) = f_j + \frac{\alpha_0}{f_0}u_{0j} - \frac{\alpha_0}{f_0}u_{0j} = f_j. \tag{2.52}$$

Since we now know $s_i^2 = 1$ $(i = 0, 1, 2)$, in order to confirm $(s_0 s_1)^3 = 1$ we may consider the braid relation $s_0 s_1 s_0 = s_1 s_0 s_1$. Let us compute $s_0 s_1 s_0(f_j)$:

$$\begin{aligned} s_0 s_1 s_0(f_j) &= s_0 s_1\left(f_j + \frac{\alpha_0}{f_0}u_{0j}\right) \\ &= s_0\left(f_j + \frac{\alpha_1}{f_1}u_{1j} + \frac{(\alpha_0+\alpha_1)f_1}{f_0 f_1 - \alpha_1}u_{0j}\right) \\ &= f_j + \frac{\alpha_0}{f_0}u_{0j} + \frac{(\alpha_0+\alpha_1)f_0}{f_0 f_1 + \alpha_0}u_{1j} + \frac{\alpha_1(f_0 f_1 + \alpha_0)}{f_0(f_0 f_1 - \alpha_1)}u_{0j} \\ &= f_j + \frac{(\alpha_0+\alpha_1)f_1}{f_0 f_1 - \alpha_1}u_{0j} + \frac{(\alpha_0+\alpha_1)f_0}{f_0 f_1 + \alpha_0}u_{1j}. \end{aligned} \tag{2.53}$$

By carrying out a similar computation again we can see that $s_1 s_0 s_1(f_j)$ gives the same result. Since we already know that s_0, s_1, s_2, π are Bäcklund transformations for the symmetric form, these transformations should commute with the derivative $'$. Although it is not necessary from a theoretical point of view, it would be meaningful to try verifying this: namely,

$$s_i(f_j)' = s_i(f_j'), \quad \pi(f_j)' = \pi(f_j') \quad (i, j = 0, 1, 2). \tag{2.54}$$

You can now see *how nicely the symmetric form matches the action of* s_i. (You might also find it a clever idea to use the Demazure operators for computing the action of s_i. ⇒ Appendix 8: Demazure operators)

Now that we have the table of Bäcklund transformations at hand, you may wish to start computing Bäcklund transformations of the particular solutions discussed in Section 2.2. Instead we will postpone this computation to Chapter 4; in the next chapter we will first investigate the τ-functions for the symmetric form.

2.5. The symmetric form of $P_{\rm II}$

In this section we formulate a symmetric form for $P_{\rm II}$, similar to the case of $P_{\rm IV}$, in order to understand better the structure of Bäcklund transformations and particular solutions. Recall that in terms of the Hamiltonian representation we have

$$H_{\rm II}: \qquad q' = p - q^2 - \frac{t}{2}, \quad p' = 2qp + b. \tag{2.55}$$

Riccati solutions appear when the parameter b takes integer values. Taking $b = 0$ and $b = 1$ as reference points, we set

$$\alpha_1 = b, \quad \alpha_0 = 1 - b \quad (\alpha_0 + \alpha_1 = 1). \tag{2.56}$$

When $b = 0$ the Riccati solutions are obtained by the specialization $p = 0$. In view of this fact we set

$$f_1 = p, \quad f_0 = r(p) = -p + 2q^2 + t. \tag{2.57}$$

Then the equations for f_0, f_1, q are expressed as

$$\begin{cases} f_0' = -2qf_0 + \alpha_0, \\ f_1' = 2qf_1 + \alpha_1, \\ q' = \dfrac{1}{2}(f_1 - f_0), \end{cases} \tag{2.58}$$

which we call the symmetric form of $P_{\rm II}$. Although this system of equations is of third order, we have

$$(f_0 + f_1)' = 2q(f_1 - f_0) + 1 = 4qq' + 1 = (2q^2 + t)', \tag{2.59}$$

and hence

$$f_0 + f_1 - 2q^2 = t + c \tag{2.60}$$

with c being the integration constant. If we normalize the constant c to be 0, by using a parallel translation with respect to the t variable, then our symmetric form of $P_{\rm II}$ as defined above is equivalent to the original $H_{\rm II}$.

In this form it is obvious that when $\alpha_0 = 0$ ($b = 1$) or $\alpha_1 = 0$ ($b = 0$) the system reduces to a Riccati equation for q by the specialization $f_0 = 0$ or $f_1 = 0$, respectively. Notice that the Bäcklund transformation r for $P_{\rm II}$ corresponds to π for the symmetric form of $P_{\rm IV}$. Let us use the same symbol $\pi = r$ to refer to the Bäcklund transformation r for the symmetric $P_{\rm II}$:

$$\pi(\alpha_0) = \alpha_1, \quad \pi(\alpha_1) = \alpha_0; \quad \pi(f_0) = f_1, \quad \pi(f_1) = f_0, \quad \pi(q) = -q. \tag{2.61}$$

This represents the fact that, together with the sign change of q, (2.58) is invariant under the operation of interchanging α_0, α_1 and f_0, f_1 simultaneously. Furthermore, the rational solution $q = 0$ arises by imposing the invariance under π, namely, $\alpha_0 = \alpha_1 = \frac{1}{2}$ and $f_0 = f_1$.

If we set $s_1 = s$ and $s_0 = rsr$, then these are the Bäcklund transformations corresponding to the reflections with respect to $\alpha_1 = 0$ and $\alpha_0 = 0$ respectively. Thus we obtain a table of Bäcklund transformations for the symmetric form of $P_{\rm II}$ as in Table 2.2. In this context the group $W = \langle s_0, s_1 \rangle$ is the affine Weyl group of type $A_1^{(1)}$. A bigger group $\widetilde{W} = \langle s_0, s_1, \pi \rangle$ acts on $P_{\rm II}$ as a group of Bäcklund transformations; $\widetilde{W} = \langle s_0, s_1, \pi \rangle$ is an extension of $W = \langle s_0, s_1 \rangle$ by an element π

TABLE 2.2. Bäcklund Transformations for the Symmetric Form of $P_{\rm II}$

	α_0	α_1	f_0	f_1	q
s_0	$-\alpha_0$	$\alpha_1+2\alpha_0$	f_0	$f_1-\frac{4\alpha_0 q}{f_0}+\frac{2\alpha_0^2}{f_0^2}$	$q-\frac{\alpha_0}{f_0}$
s_1	$\alpha_0+2\alpha_1$	$-\alpha_1$	$f_0+\frac{4\alpha_1 q}{f_1}+\frac{2\alpha_1^2}{f_1^2}$	f_1	$q+\frac{\alpha_1}{f_1}$
π	α_1	α_0	f_1	f_0	$-q$

of order 2 that interchanges s_0, s_1. The Cartan matrix and the Dynkin diagram of type $A_1^{(1)}$ are given as follows:

$$A=\begin{bmatrix} 2 & -2 \\ -2 & 2 \end{bmatrix}, \qquad 0\circ \Longleftrightarrow \circ\, 1\,. \tag{2.62}$$

CHAPTER 3

τ-Functions

In this chapter we introduce the τ-functions for the symmetric form by means of the Poisson structure. Furthermore we investigate basic properties of the τ-functions, including their differential equations and Bäcklund transformations.

3.1. Poisson structure and the Hamiltonian system

Keeping the notation of the previous chapter, we consider the symmetric form of $P_{\rm IV}$:

$$
\left\{
\begin{aligned}
f_0' &= f_0(f_1 - f_2) + \alpha_0, \\
f_1' &= f_1(f_2 - f_0) + \alpha_1, \\
f_2' &= f_2(f_0 - f_1) + \alpha_2,
\end{aligned}
\right.
\qquad
\begin{aligned}
&\alpha_0 + \alpha_1 + \alpha_2 = 1, \\
&f_0 + f_1 + f_2 = t.
\end{aligned}
\tag{3.1}
$$

This system of differential equations can also be written as a Hamiltonian system, similar to the case of $P_{\rm II}$. When we discussed $P_{\rm II}$, we first considered the Hamiltonian system and then introduced the Poisson bracket so that the variables (q, p) should give the canonical coordinate system. This time we proceed in the opposite way; we first define the Poisson bracket and then derive a representation of $P_{\rm IV}$ as a Hamiltonian system. In what follows, by a *Poisson bracket* we mean an operation of attaching a function $\{f, g\}$ to each pair of functions f, g in some space of functions, imposing the following conditions:

(1) The bracket $\{\,,\,\}$ is bilinear and skew-symmetric:

$$\{f, f\} = 0, \quad \{g, f\} = -\{f, g\}.$$

(2) The bracket $\{\,,\,\}$ satisfies the Leibniz rule in each of the two arguments:

$$\{fg, h\} = \{f, h\}g + f\{g, h\}, \quad \{f, gh\} = \{f, g\}h + g\{f, h\}.$$

(3) The bracket $\{\,,\,\}$ satisfies the Jacobi rule:

$$\{f, \{g, h\}\} + \{g, \{h, f\}\} + \{h, \{f, g\}\} = 0.$$

Recall that in Section 2.4 we have used the skew-symmetric matrix

$$
U = (u_{ij})_{i,j=0}^{2} = \begin{bmatrix} 0 & 1 & -1 \\ -1 & 0 & 1 \\ 1 & -1 & 0 \end{bmatrix}
\tag{3.2}
$$

for describing the Bäcklund transformations of the symmetric form. This matrix can be thought of as specifying the orientation of the Dynkin diagram (a way to assign an arrow to each edge connecting ∘'s). Namely, whenever $u_{ij} = 1$ we draw an arrow from the node named i towards the one named j. When $u_{ij} = -1$ we already have an arrow from the node j to i since $u_{ij} = -u_{ji}$. In this particular case the matrix U above corresponds to the following orientation of the triangle:

(3.3)

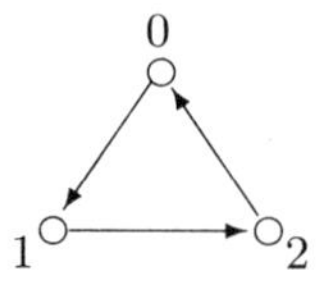

By using this matrix U we define the Poisson bracket among the variables f_0, f_1, f_2 by

$$\{f_i, f_j\} = u_{ij} \qquad (i, j = 0, 1, 2). \tag{3.4}$$

Since $\{f_i, f_i\} = 0$ and $\{f_j, f_i\} = -\{f_i, f_j\}$, the bracket is determined by the formula

$$\{f_0, f_1\} = \{f_1, f_2\} = \{f_2, f_0\} = 1. \tag{3.5}$$

In other words, the Poisson bracket takes the value 1 if the pair of variables f_j is combined according to an arrow attached to the Dynkin diagram. From this information, by using the rules mentioned above one can determine the Poisson bracket for any pair of functions expressed in terms of f_0, f_1, f_2. One can also define the Poisson bracket directly for any functions φ, ψ in f_0, f_1, f_2:

$$\{\varphi, \psi\} = \sum_{i,j=0}^{2} \frac{\partial \varphi}{\partial f_i} u_{i,j} \frac{\partial \psi}{\partial f_j}. \tag{3.6}$$

PROPOSITION 3.1. *The bracket $\{\ ,\ \}$ for functions in f_0, f_1, f_2 defined by (3.6) is a Poisson bracket. Furthermore, it is invariant under the action of $\widetilde{W} = \langle\, s_0, s_1, s_2, \pi \,\rangle$, namely,*

$$w(\{\varphi, \psi\}) = \{w(\varphi), w(\psi)\} \quad (w \in \widetilde{W}). \tag{3.7}$$

One can directly verify that $\{\ ,\}$ satisfies the three conditions of the Poisson bracket. Note also that (3.6) implies $\{f_i, f_j\} = u_{ij}$. The invariance under the action of $\widetilde{W}$ can be proved by computation as follows. We have only to check the case where $\varphi = f_i, \psi = f_j$. Since $s_k(\{f_i, f_j\}) = s_k(u_{ij}) = u_{ij}$ we need to prove $\{s_k(f_i), s_k(f_j)\} = u_{ij}$. In fact we have

$$\begin{aligned} \{s_k(f_i), s_k(f_j)\} &= \left\{ f_i + \frac{\alpha_k}{f_k} u_{ki}, f_j + \frac{\alpha_k}{f_k} u_{kj} \right\} \\ &= \{f_i, f_j\} - \frac{\alpha_k}{f_k^2} u_{ki} \{f_k, f_j\} - \frac{\alpha_k}{f_k^2} u_{kj} \{f_i, f_k\} \\ &= u_{ij} - \frac{\alpha_k}{f_k^2} (u_{ki} u_{kj} + u_{kj} u_{ik}) = u_{ij}. \end{aligned} \tag{3.8}$$

For a general Poisson bracket we will also define a *canonical transformation* to be a transformation that preserves the Poisson bracket. In this terminology the proposition mentioned above implies that each element of $\widetilde{W}$ gives rise to a birational canonical transformation with respect to the variables f_0, f_1, f_2. We remark that the Poisson bracket of a general function φ with f_j is given by the action of a vector field as follows:

$$\{\varphi, f_j\} = \sum_{i=0}^{2} \frac{\partial \varphi}{\partial f_i} u_{i,j} = \frac{\partial \varphi}{\partial f_{j-1}} - \frac{\partial \varphi}{\partial f_{j+1}}. \tag{3.9}$$

In particular we have

$$\{\varphi, t\} = \{\varphi, f_0 + f_1 + f_2\} = 0. \tag{3.10}$$

Namely, any function depending only on t behaves *like a constant* under this Poisson bracket.

We now consider the following polynomial of degree 3:

$$H = f_0 f_1 f_2 + b_0 f_0 + b_1 f_1 + b_2 f_2, \tag{3.11}$$

where b_0, b_1, b_2 are parameters. The Poisson brackets of this H with f_j's are computed as follows:

$$\begin{aligned} \{H, f_0\} &= f_0(f_1 - f_2) + (b_2 - b_1), \\ \{H, f_1\} &= f_1(f_2 - f_0) + (b_0 - b_2), \\ \{H, f_2\} &= f_2(f_0 - f_1) + (b_1 - b_0). \end{aligned} \tag{3.12}$$

Observe that these formulas are very close to the right-hand sides of the symmetric form. If we could take b_0, b_1, b_2 so that

$$b_2 - b_1 = \alpha_0, \quad b_0 - b_2 = \alpha_1, \quad b_1 - b_0 = \alpha_2, \tag{3.13}$$

then our symmetric form would be expressed as

$$f_j' = \{H, f_j\} \quad (j = 0, 1, 2). \tag{3.14}$$

But this is impossible under the condition $\alpha_0 + \alpha_1 + \alpha_2 = 1 \neq 0$. So we distinguish the index 0 from the other two and try to find b_0, b_1, b_2 such that

$$b_2 - b_1 = \alpha_0 - 1, \quad b_0 - b_2 = \alpha_1, \quad b_1 - b_0 = \alpha_2. \tag{3.15}$$

From the one-dimensional solution space of this equation we pick up the following solution by the normalization $b_0 + b_1 + b_2 = 0$:

$$b_0 = \frac{1}{3}(\alpha_1 - \alpha_2), \quad b_1 = \frac{1}{3}(\alpha_1 + 2\alpha_2), \quad b_2 = -\frac{1}{3}(2\alpha_1 + \alpha_2). \tag{3.16}$$

The argument above implies the following proposition.

PROPOSITION 3.2. *The symmetric form of $P_{\rm IV}$ is represented as*

$$\left\{ \begin{aligned} f_0' &= \{H, f_0\} + 1, \\ f_1' &= \{H, f_1\}, \\ f_2' &= \{H, f_2\} \end{aligned} \right. \tag{3.17}$$

in terms of the Poisson bracket, where

$$H = f_0 f_1 f_2 + \frac{1}{3}(\alpha_1 - \alpha_2) f_0 + \frac{1}{3}(\alpha_1 + 2\alpha_2) f_1 - \frac{1}{3}(2\alpha_1 + \alpha_2) f_2. \tag{3.18}$$

For a general function φ of f_0, f_1, f_2 we have

$$\varphi' = \sum_{j=0}^{2} \frac{\partial \varphi}{\partial f_j} f_j' = \sum_{j=0}^{2} \frac{\partial \varphi}{\partial f_j}\{H, f_j\} + \frac{\partial \varphi}{\partial f_0} = \{H, \varphi\} + \frac{\partial \varphi}{\partial f_0}. \tag{3.19}$$

By now we have introduced the Poisson structure in the 3-dimensional space with coordinate system (f_0, f_1, f_2). In view of $\{f_1, f_2\} = 1$ we set

$$p = f_1, \quad q = f_2, \quad t = f_0 + f_1 + f_2. \tag{3.20}$$

By doing so we clearly obtain a new coordinate system (q, p, t) such that

$$\{p, q\} = 1, \quad \{q, t\} = \{p, t\} = 0. \tag{3.21}$$

In terms of these coordinates the same Poisson bracket can be expressed as

$$\{\varphi, \psi\} = \frac{\partial \varphi}{\partial p}\frac{\partial \psi}{\partial q} - \frac{\partial \varphi}{\partial q}\frac{\partial \psi}{\partial p}. \tag{3.22}$$

Hence we have

$$q' = \{H, q\} = \frac{\partial H}{\partial p}, \quad p' = \{H, p\} = -\frac{\partial H}{\partial q}. \tag{3.23}$$

Namely, we see that the symmetric form is expressed as a Hamiltonian system. Since

$$f_0 = t - q - p, \quad f_1 = p, \quad f_2 = q, \tag{3.24}$$

the Hamiltonian is computed as

$$H = (t - q - p)pq + \alpha_2 p - \alpha_1 q + \frac{1}{3}(\alpha_1 - \alpha_2)t. \tag{3.25}$$

PROPOSITION 3.3. *The symmetric form of P_{IV} is equivalent to the Hamiltonian system with Hamiltonian*

$$H = (t - q - p)pq + \alpha_2 p - \alpha_1 q + \frac{1}{3}(\alpha_1 - \alpha_2)t, \tag{3.26}$$

namely,

$$\begin{cases} q' = q(t - q - 2p) + \alpha_2, \\ p' = p(2q + p - t) + \alpha_1. \end{cases} \tag{3.27}$$

In this section, we first introduced the Poisson structure by using the matrix $U = (u_{i,j})_{i,j=0}^2$ and then deduced the expression of the symmetric form of P_{IV} as a Hamiltonian system. If we regard the Poisson bracket as the primary object, then the matrix U is nothing but the table of the Poisson brackets:

(3.28)

$\{,\}$	f_0	f_1	f_2
f_0	0	1	−1
f_1	−1	0	1
f_2	1	−1	0

Furthermore, we see that the Bäcklund transformations s_0, s_1, s_2 are expressed as follows in terms of the Poisson bracket:

$$s_i(f_j) = f_j + \frac{\alpha_i}{f_i}\{f_i, f_j\} \qquad (i, j = 0, 1, 2). \tag{3.29}$$

3.2. τ-Functions and their differential equations

As we mentioned in Section 1.1, the *τ-function* for a Hamiltonian system is introduced as a variable such that

$$H = \frac{d}{dt} \log \tau = \frac{\tau'}{\tau}, \tag{3.30}$$

by means of the Hamiltonian H. I want to remind you here that there is ambiguity in choosing a Hamiltonian. If one modifies H by adding a function depending only on $t = f_0 + f_1 + f_2$, say $c(t)$, it still remains a Hamiltonian for the same differential equation. Note that in such a case the τ-function is multiplied by $e^{\int c(t)dt}$. Although we will use a particular Hamiltonian H in the following argument, it would be meaningful to keep in mind that it involves some freedom as explained above.

In defining the Hamiltonian H in (3.18) we gave a special role to the index 0. For this reason the three indices $0, 1, 2$ are no longer equal in the expression of H. In order to recover the symmetry we set $h_0 = H$ and introduce two more Hamiltonians $h_1 = \pi(h_0)$, $h_2 = \pi(h_1)$ by rotating the indices:

$$\begin{aligned} h_0 &= f_0 f_1 f_2 + \frac{\alpha_1 - \alpha_2}{3} f_0 + \frac{\alpha_1 + 2\alpha_2}{3} f_1 - \frac{2\alpha_1 + \alpha_2}{3} f_2, \\ h_1 &= f_0 f_1 f_2 - \frac{2\alpha_2 + \alpha_0}{3} f_0 + \frac{\alpha_2 - \alpha_0}{3} f_1 + \frac{\alpha_2 + 2\alpha_0}{3} f_2, \\ h_2 &= f_0 f_1 f_2 + \frac{\alpha_0 + 2\alpha_1}{3} f_0 - \frac{2\alpha_0 + \alpha_1}{3} f_1 + \frac{\alpha_0 - \alpha_1}{3} f_2. \end{aligned} \tag{3.31}$$

Accordingly, we introduce three τ-functions τ_0, τ_1, τ_2 and deal with them equally:

$$h_0 = \frac{\tau_0'}{\tau_0}, \quad h_1 = \frac{\tau_1'}{\tau_1}, \quad h_2 = \frac{\tau_2'}{\tau_2}. \tag{3.32}$$

Note that each τ_i is determined only up to multiplication by a constant.

Notice that the three Hamiltonians h_0, h_1, h_2 have a common component $f_0 f_1 f_2$ of degree 3, and that the differences appear only in the components of degree 1. By using $\alpha_0 + \alpha_1 + \alpha_2 = 1$ the difference of h_1 and h_2 is computed as

$$h_2 - h_1 = \frac{2}{3} f_0 - \frac{1}{3} f_1 - \frac{1}{3} f_2 = f_0 - \frac{1}{3}(f_0 + f_1 + f_2) = f_0 - \frac{t}{3}. \tag{3.33}$$

Hence we have $f_0 = h_2 - h_1 + \frac{t}{3}$. This provides the formula

$$f_0 = \frac{\tau_2'}{\tau_2} - \frac{\tau_1'}{\tau_1} + \frac{t}{3}, \tag{3.34}$$

representing f_0 in terms of the τ-functions. By the symmetry with respect to the indices, we have similar formulas for f_1, f_2 as well.

PROPOSITION 3.4. *The variables f_j are recovered from the τ-functions by the following formulas:*

$$
\begin{aligned}
f_0 &= h_2 - h_1 + \frac{t}{3} = \frac{\tau_2'}{\tau_2} - \frac{\tau_1'}{\tau_1} + \frac{t}{3},\\
f_1 &= h_0 - h_2 + \frac{t}{3} = \frac{\tau_0'}{\tau_0} - \frac{\tau_2'}{\tau_2} + \frac{t}{3},\\
f_2 &= h_1 - h_0 + \frac{t}{3} = \frac{\tau_1'}{\tau_1} - \frac{\tau_0'}{\tau_0} + \frac{t}{3}.
\end{aligned}
\tag{3.35}
$$

From this proposition we get

$$
\begin{aligned}
f_1 - f_2 &= 2\,h_0 - h_1 - h_2,\\
f_2 - f_0 &= -h_0 + 2h_1 - h_2,\\
f_0 - f_1 &= -h_0 - h_1 + 2h_2.
\end{aligned}
\tag{3.36}
$$

Hence the differential equations for f_0, f_1, f_2 are expressed as

$$
\begin{aligned}
f_0' &= f_0(2\,h_0 - h_1 - h_2) + \alpha_0,\\
f_1' &= f_1(-h_0 + 2h_1 - h_2) + \alpha_1,\\
f_2' &= f_2(-h_0 - h_1 + 2h_2) + \alpha_2.
\end{aligned}
\tag{3.37}
$$

(This is the Cartan matrix!)

We now deduce the differential equations to be satisfied by the new dependent variables τ_0, τ_1, τ_2. For this purpose we need to compute the derivative of $h_0 = \tau_0'/\tau_0$. Since h_0 is a polynomial of f_0, f_1, f_2, we can apply (3.19) to get

$$
h_0' = \{h_0, h_0\} + \frac{\partial h_0}{\partial f_0} = \frac{\partial h_0}{\partial f_0} = f_1 f_2 + \frac{\alpha_1 - \alpha_2}{3}.
\tag{3.38}
$$

We have only to rotate the indices by π to get the corresponding formulas for h_1, h_2:

$$
\begin{aligned}
h_0' &= f_1 f_2 + \frac{1}{3}(\alpha_1 - \alpha_2),\\
h_1' &= f_2 f_0 + \frac{1}{3}(\alpha_2 - \alpha_0),\\
h_2' &= f_0 f_1 + \frac{1}{3}(\alpha_0 - \alpha_1).
\end{aligned}
\tag{3.39}
$$

Taking the sum of the first two, we compute

$$
\begin{aligned}
h_0' + h_1' &= f_2(f_0 + f_1) - \frac{1}{3}(\alpha_0 - \alpha_1)\\
&= f_2(t - f_2) - \frac{1}{3}(\alpha_0 - \alpha_1)\\
&= \left(h_1 - h_0 + \frac{t}{3}\right)\left(h_0 - h_1 + \frac{2t}{3}\right) - \frac{1}{3}(\alpha_0 - \alpha_1),
\end{aligned}
\tag{3.40}
$$

and hence

$$
(h_0 + h_1)' + (h_0 - h_1)^2 + \frac{t}{3}(h_0 - h_1) - \frac{2t^2}{9} + \frac{\alpha_0 - \alpha_1}{3} = 0.
\tag{3.41}
$$

By using $h_j = \tau_j'/\tau_j$ this formula can be rewritten into the following equation for the τ-functions:

$$\tau_0''\tau_1 - 2\tau_0'\tau_1' + \tau_0\tau_1'' + \frac{t}{3}(\tau_0'\tau_1 - \tau_0\tau_1') - \left(\frac{2t^2}{9} - \frac{\alpha_0-\alpha_1}{3}\right)\tau_0\tau_1 = 0. \tag{3.42}$$

In this way we have obtained a *bilinear* differential equation for the τ-functions. By saying "bilinear" we mean that it is linear with respect to the individual τ_i $(i = 0, 1)$, although the equation as a whole is quadratic. Thanks to the symmetry with respect to the indices we obtain similar bilinear equations for the pairs (τ_1, τ_2) and (τ_2, τ_0).

This class of bilinear differential equations can be expressed neatly in terms of the *Hirota derivatives.* For a given pair of functions f and g, the Hirota derivatives are defined by

$$D_t\, f \cdot g = f'\, g - f\, g', \quad D_t^2\, f \cdot g = f''\, g - 2f'\, g' + f\, g'', \ldots \tag{3.43}$$

by combining the derivatives of f and g according to the Leibniz rule with alternating signs. For example, the formulas (3.35) that recover f_j from the τ-functions are expressed as

$$f_j = \frac{1}{\tau_{j-1}\tau_{j+1}}\left(D_t + \frac{t}{3}\right)\tau_{j-1}\cdot\tau_{j+1} \quad (j = 0, 1, 2). \tag{3.44}$$

Remark: The symbol " · " inserted in between f and g is *not* meant to be the multiplication; the expression $D_t f \cdot g$ as a whole denotes a function. For details ⇒ Appendix 9: Hirota's bilinear operators.

The bilinear equation (3.42) above can also be expressed by means of the Hirota derivatives.

THEOREM 3.5. *For the symmetric form of P_{IV} the differential equations for the three τ-functions τ_0, τ_1, τ_2 are given by the following bilinear differential equations of Hirota type*:

$$\begin{aligned}
&\left(D_t^2 + \frac{t}{3}D_t - \frac{2}{9}t^2 + \frac{\alpha_0-\alpha_1}{3}\right)\tau_0\cdot\tau_1 = 0,\\
&\left(D_t^2 + \frac{t}{3}D_t - \frac{2}{9}t^2 + \frac{\alpha_1-\alpha_2}{3}\right)\tau_1\cdot\tau_2 = 0,\\
&\left(D_t^2 + \frac{t}{3}D_t - \frac{2}{9}t^2 + \frac{\alpha_2-\alpha_0}{3}\right)\tau_2\cdot\tau_0 = 0.
\end{aligned} \tag{3.45}$$

Supposing that the three τ-functions τ_0, τ_1, τ_2 satisfy the Hirota bilinear equations above, define the variables f_0, f_1, f_2 by

$$f_j = h_{j-1} - h_{j+1} + \frac{t}{3}, \quad h_j = \frac{\tau_j'}{\tau_j} \quad (j = 0, 1, 2). \tag{3.46}$$

We now verify that the f variables defined in this way certainly solve the symmetric form of P_{IV}. It is obvious that they satisfy the normalization condition $f_0+f_1+f_2 =$

t. The first two equations in (3.45) imply the relations

$$(3.47)\qquad \begin{aligned}(h_0+h_1)'+(h_0-h_1)^2+\frac{t}{3}(h_0-h_1)-\frac{2t^2}{9}+\frac{\alpha_0-\alpha_1}{3}&=0,\\(h_1+h_2)'+(h_1-h_2)^2+\frac{t}{3}(h_1-h_2)-\frac{2t^2}{9}+\frac{\alpha_1-\alpha_2}{3}&=0\end{aligned}$$

for $h_j=\tau_j'/\tau_j$ $(j=0,1,2)$. Taking the difference of these two formulas, we obtain

$$(3.48)\qquad (h_0-h_2)'-\left(h_0-h_2+\frac{t}{3}\right)(-h_0+2h_1-h_2)+\frac{1}{3}-\alpha_1=0,$$

namely,

$$(3.49)\qquad \left(h_0-h_2+\frac{t}{3}\right)'=\left(h_0-h_2+\frac{t}{3}\right)(-h_0+2h_1-h_2)+\alpha_1.$$

Since $f_1=h_0-h_2+\frac{t}{3}$ and $f_2-f_0=-h_0+2h_1-h_2$, this implies

$$(3.50)\qquad f_1'=f_1(f_2-f_0)+\alpha_1.$$

The corresponding equations for f_0 and f_2 are obtained similarly.

In this section, we introduced three τ-functions for the symmetric form of P_{IV} and derived the Hirota bilinear equations for them. We will show below explicit examples of τ-functions corresponding to the special solutions discussed in Section 2.2.

If we take the rational solution

$$(3.51)\qquad (\alpha_0,\alpha_1,\alpha_2;f_0,f_1,f_2)=\left(\frac{1}{3},\frac{1}{3},\frac{1}{3};\frac{t}{3},\frac{t}{3},\frac{t}{3}\right)$$

at the barycenter of the fundamental triangle, the corresponding Hamiltonians are computed as

$$(3.52)\qquad h_0=f_0f_1f_2+\frac{\alpha_1-\alpha_2}{3}f_0+\frac{\alpha_1+2\alpha_2}{3}f_1-\frac{2\alpha_1+\alpha_2}{3}f_2=\frac{t^3}{27}.$$

Hence we have

$$(3.53)\qquad (\tau_0,\tau_1,\tau_2)=\left(c_0\exp\left(\frac{t^4}{108}\right),c_1\exp\left(\frac{t^4}{108}\right),c_2\exp\left(\frac{t^4}{108}\right)\right),$$

where c_0, c_1, $c_2\in\mathbb{C}^*=\mathbb{C}\setminus\{0\}$. (Each τ_i is determined up to a constant multiple.)

We next consider the solutions of Riccati type along $\alpha_0=0$. Let $u=\varphi(t)$ be the general solution of the differential equation of Hermite

$$(3.54)\qquad u''-t\,u'-\alpha_1u=0.$$

Then the solutions in question are expressed as

$$(3.55)\qquad (\alpha_0,\alpha_1,\alpha_2;f_0,f_1,f_2)=\left(0,\alpha_1,\alpha_2;0,\frac{\varphi'(t)}{\varphi(t)},t-\frac{\varphi'(t)}{\varphi(t)}\right).$$

In the same way as in the previous case the τ-functions are determined by computing the Hamiltonians h_0, h_1, h_2. The answer for this case is given by

$$(3.56)\quad (\tau_0,\tau_1,\tau_2)=\left(c_0\exp\left(-\frac{\alpha_1+1}{6}t^2\right)\varphi(t),c_1\exp\left(\frac{\alpha_2}{6}t^2\right),c_2\exp\left(-\frac{\alpha_1}{6}t^2\right)\right).$$

REMARK 3.6. (1) If we modify the Hamiltonians and the τ-functions by setting

$$\widehat{h}_j = h_j - \frac{t^3}{27}, \quad \widehat{\tau}_j = \exp\left(-\frac{t^4}{108}\right)\tau_j, \tag{3.57}$$

the differential equations (3.45) for the τ-functions become somewhat simpler:

$$\left(D_t^2 + \frac{t}{3}D_t + \frac{\alpha_j - \alpha_{j+1}}{3}\right)\widehat{\tau}_j \cdot \widehat{\tau}_{j+1} = 0 \qquad (j = 0, 1, 2). \tag{3.58}$$

This modification has the effect of transforming the τ-functions for the rational solution at the barycenter into constants. In fact, when $(\alpha_0, \alpha_1, \alpha_2) = (\frac{1}{3}, \frac{1}{3}, \frac{1}{3})$ the triple $(\widehat{\tau}_0, \widehat{\tau}_1, \widehat{\tau}_2) = (c_0, c_1, c_2)$ obviously solves the bilinear equations for $\widehat{\tau}_j$ above.
(2) The bilinear equations (3.45) are invariant under the transformation that replaces τ_j simultaneously by $\exp(ct)\tau_j$, c being a constant. Consider the variables f_j and h_j defined by (3.46) from the τ-functions. Then the transformation $\tau_j \to \exp(ct)\tau_j$ leaves f_j unchanged, but gives rise to a change $h_j \to h_j + c$ in the Hamiltonians. For this reason $h_j = \tau_j'/\tau_j$ may differ by a constant from the one computed by using (3.31).

3.3. Bäcklund transformations of the τ-functions

In Section 2.4 we investigated the action of the affine Weyl group on the variables f_0, f_1, f_2 of the symmetric form. Is it possible to find the corresponding Bäcklund transformations on the level of τ-functions as well? In order to answer this question, in view of $h_j = (\log \tau_j)'$ we need to understand how the Hamiltonians behave under the Bäcklund transformations.

LEMMA 3.7. *For all* $i, j = 0, 1, 2$, *one has*

$$s_i(h_j) = h_j \quad (i \neq j), \quad s_j(h_j) = h_j + \frac{\alpha_j}{f_j}. \tag{3.59}$$

We omit the proof of this lemma since it can be verified by direct computation. Note that the Hamiltonian has a characteristic property; h_0 is invariant with respect to s_1 and s_2, and responds only to s_0 with the same index.

Combining the lemma above with the differential equations for f_j, we will be able to infer how to define the Bäcklund transformations for the τ-functions. We consider the two equations

$$\frac{f_0'}{f_0} = 2h_0 - h_1 - h_2 + \frac{\alpha_0}{f_0}, \quad s_0(h_0) = h_0 + \frac{\alpha_0}{f_0}, \tag{3.60}$$

one obtained from (3.37) by division by f_0, and the other taken from the lemma above. From these we obtain

$$\frac{f_0'}{f_0} = h_0 + s_0(h_0) - h_2 - h_1 \tag{3.61}$$

by eliminating α_0/f_0. If the Bäcklund transformation s_0 makes sense on the level of the τ's as well, then it must commute with the derivative. Hence we should have

$$s_0(h_0) = s_0\left(\frac{\tau_0'}{\tau_0}\right) = \frac{s_0(\tau_0')}{s_0(\tau_0)} = \frac{s_0(\tau_0)'}{s_0(\tau_0)}. \tag{3.62}$$

Furthermore we should get

$$\frac{f_0'}{f_0} = \frac{\tau_0'}{\tau_0} + \frac{s_0(\tau_0)'}{s_0(\tau_0)} - \frac{\tau_2'}{\tau_2} - \frac{\tau_1'}{\tau_1}, \tag{3.63}$$

or equivalently

$$(\log f_0)' = \left(\log \frac{\tau_0 s_0(\tau_0)}{\tau_2\,\tau_1}\right)'. \tag{3.64}$$

This implies

$$f_0 = c_0\,\frac{\tau_0 s_0(\tau_0)}{\tau_2\,\tau_1} \qquad \text{(for some constant } c_0). \tag{3.65}$$

The argument above implies that the Bäcklund transformations s_j should be defined by the formulas

$$f_j = c_j\,\frac{\tau_j\,s_j(\tau_j)}{\tau_{j-1}\tau_{j+1}}, \quad \text{i.e.} \quad s_j(\tau_j) = \frac{f_j}{c_j}\,\frac{\tau_{j-1}\tau_{j+1}}{\tau_j} \qquad (j = 0, 1, 2). \tag{3.66}$$

It is a subtle question how to fix the constants c_j, since they may depend on the parameters α_0, α_1, α_2. While being aware of this point, we will make the simplest possible choice by setting $c_0 = c_1 = c_2 = 1$.

THEOREM 3.8. *Define the transformations s_0, s_1, s_2, π for the τ-functions by*

$$s_i(\tau_j) = \tau_j \quad (i \neq j), \quad s_j(\tau_j) = f_j\,\frac{\tau_{j-1}\tau_{j+1}}{\tau_j}, \quad \pi(\tau_j) = \tau_{j+1} \tag{3.67}$$

for $i, j = 0, 1, 2$. Then these provide Bäcklund transformations for the τ-functions. Furthermore they satisfy the relations

$$s_j^2 = 1, \quad (s_j s_{j+1})^3 = 1; \quad \pi^3 = 1, \quad \pi s_j = s_{j+1}\pi \quad (j = 0, 1, 2) \tag{3.68}$$

with respect to all the variables including the τ-functions.

Namely the action of the extended affine Weyl group of type $A_2^{(1)}$ is extended to the level of Bäcklund transformations for τ-functions. Suppose for instance that a triple (τ_0, τ_1, τ_2) of τ-functions satisfies the three bilinear equations (3.45) of Theorem 3.5 for some parameters $\alpha_0, \alpha_1, \alpha_2$ $(\alpha_0 + \alpha_1 + \alpha_2 = 1)$. If we set

$$\widetilde{\tau}_0 = f_0\frac{\tau_2\tau_1}{\tau_0}, \quad \widetilde{\tau}_1 = \tau_1, \quad \widetilde{\tau}_2 = \tau_2 \tag{3.69}$$

by using the formulas for s_0, then the new variables $\widetilde{\tau}_j$ satisfy the differential equations of the same form with parameters replaced by their s_0-transforms

$$\widetilde{\alpha}_0 = -\alpha_0, \quad \widetilde{\alpha}_1 = \alpha_1 + \alpha_0, \quad \widetilde{\alpha}_2 = \alpha_2 + \alpha_0. \tag{3.70}$$

We remark that the definition of $s_j(f_j)$ mentioned above can be understood as the multiplicative formula

$$f_j = \frac{\tau_j\,s_j(\tau_j)}{\tau_{j-1}\tau_{j+1}} = \frac{\tau_j\,s_j(\tau_j)}{\prod_{i\neq j}\tau_i^{-a_{ij}}} \qquad (j = 0, 1, 2) \tag{3.71}$$

expressing f_j in terms of the τ-functions. Also by using the previous (3.35) or (3.44) the s_j-transform of τ_j can be represented as

$$
\begin{aligned}
s_j(\tau_j) &= \frac{1}{\tau_j}\left(\tau'_{j-1}\tau_{j+1} - \tau_{j-1}\tau'_{j+1} + \frac{t}{3}\tau_{j-1}\tau_{j+1}\right) \\
&= \frac{1}{\tau_j}\left(D_t + \frac{t}{3}\right)\tau_{j-1}\cdot\tau_{j+1},
\end{aligned} \tag{3.72}
$$

by means of the Hirota derivatives.

In order to prove that s_0, s_1, s_2 defined as above provide Bäcklund transformations for the τ-functions, we have only to trace their construction again. It is easy to check that π gives a Bäcklund transformation. Here we will verify that s_0, s_1, s_2 keep satisfying the relations for the affine Weyl group of type $A_2^{(1)}$. When $i \neq j$ we clearly have $s_i^2(\tau_j) = \tau_j$. Since

$$
s_i^2(\tau_i) = s_i\left(f_i\frac{\tau_{i-1}\tau_{i+1}}{\tau_i}\right) = f_i\frac{\tau_{i-1}\tau_{i+1}}{s_i(\tau_i)} = \tau_i, \tag{3.73}
$$

we have $s_i^2 = 1$ on the level of τ-functions. We next prove the braid relation $s_1s_2s_1 = s_2s_1s_2$. Both sides of this relation keep τ_0 invariant. So we compute $s_1s_2(\tau_1)$:

$$
\begin{aligned}
s_2s_1(\tau_1) &= s_2\left(f_1\frac{\tau_0\tau_2}{\tau_1}\right) = s_2(f_1)\frac{\tau_0\, s_2(\tau_2)}{\tau_1} \\
&= \left(f_1 - \frac{\alpha_2}{f_2}\right)\frac{\tau_0}{\tau_1}\, f_2\, \frac{\tau_1\,\tau_0}{\tau_2} = (f_1f_2 - \alpha_2)\frac{\tau_0^2}{\tau_2}.
\end{aligned} \tag{3.74}
$$

Hence we have

$$
s_2s_1(\tau_1) = (f_1f_2 - \alpha_2)\frac{\tau_0^2}{\tau_2}. \tag{3.75}
$$

We now look at the coefficient $f_1f_2 - \alpha_2$. Notice that

$$
\begin{aligned}
s_1(f_1f_2 - \alpha_2) &= f_1s_1(f_2) - s_1(\alpha_2) \\
&= f_1\left(f_2 + \frac{\alpha_1}{f_1}\right) - (\alpha_2 + \alpha_1) = f_1f_2 - \alpha_2,
\end{aligned} \tag{3.76}
$$

namely, $f_1f_2 - \alpha_2$ is s_1-invariant. This implies that $s_2s_1(\tau_1)$ is s_1-invariant as well. Hence we have

$$
s_1s_2s_1(\tau_1) = s_2s_1(\tau_1) = s_2s_1s_2(\tau_1). \tag{3.77}
$$

Similarly we have

$$
s_1s_2(\tau_2) = (f_1f_2 + \alpha_1)\frac{\tau_0^2}{\tau_1}. \tag{3.78}
$$

This time $f_1f_2 + \alpha_1$ is s_2-invariant, and it implies $s_1s_2s_1(\tau_2) = s_2s_1s_2(\tau_2)$. The other braid relations are obtained by the rotation of indices.

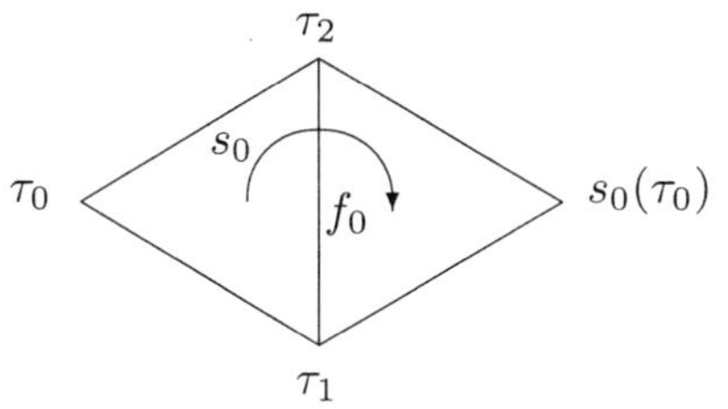

$$f_0 = \frac{\tau_0\, s_0(\tau_0)}{\tau_2\, \tau_1}$$

FIGURE 3.1. Transformation by s_0

3.4. Various relations among the τ-functions

In the previous section we have seen that the Bäcklund transformations s_0, s_1, s_2, π can be lifted to the level of τ-functions with the relations of the extended affine Weyl group preserved. As for the Bäcklund transformations of the τ-functions, the following formulas are essential:

$$\begin{aligned} s_0(\tau_0) &= f_0\, \frac{\tau_2\tau_1}{\tau_0} = \frac{1}{\tau_0}\left(D_t + \frac{t}{3}\right) \tau_2 \cdot \tau_1, \\ s_1(\tau_1) &= f_1\, \frac{\tau_0\tau_1}{\tau_0} = \frac{1}{\tau_1}\left(D_t + \frac{t}{3}\right) \tau_0 \cdot \tau_2, \\ s_2(\tau_2) &= f_2\, \frac{\tau_1\tau_2}{\tau_0} = \frac{1}{\tau_2}\left(D_t + \frac{t}{3}\right) \tau_1 \cdot \tau_0. \end{aligned} \tag{3.79}$$

These are also equivalent to the formulas

$$\begin{aligned} f_0 &= \frac{\tau_0\, s_0(\tau_0)}{\tau_2\, \tau_1} = \frac{\tau_2'}{\tau_2} - \frac{\tau_1'}{\tau_1} + \frac{t}{3}, \\ f_1 &= \frac{\tau_1\, s_1(\tau_1)}{\tau_0\, \tau_2} = \frac{\tau_0'}{\tau_0} - \frac{\tau_2'}{\tau_2} + \frac{t}{3}, \\ f_2 &= \frac{\tau_2\, s_2(\tau_2)}{\tau_1\, \tau_0} = \frac{\tau_1'}{\tau_1} - \frac{\tau_0'}{\tau_0} + \frac{t}{3} \end{aligned} \tag{3.80}$$

that recover the original variables f_0, f_1, f_2 from the τ-functions. A graphical expression like Figure 3.1 would be helpful in understanding the structure of the transformation by s_i. Similarly Figure 3.2 shows how the six τ-functions all together recover the f variables. In these figures each transformation s_i is thought of as the reflection with respect to the line to which f_i is attached; every time we apply s_i to a τ-function, we produce its mirror image, by placing a new τ-function at the corresponding symmetrical point.

By using such a symmetry structure various formulas concerning the τ-functions can be derived from our knowledge of the f variables. In this section we will present some such formulas.

3.4.1. Algebraic relations among six τ-functions.

Recalling that $f_0 + f_1 + f_2 = t$, we translate this into the language of τ-functions:

$$\frac{\tau_0 s_0(\tau_0)}{\tau_2\, \tau_1} + \frac{\tau_1 s_1(\tau_1)}{\tau_0\, \tau_2} + \frac{\tau_2 s_2(\tau_2)}{\tau_1\, \tau_0} = t. \tag{3.81}$$

Clearing the denominators, we obtain

$$\tau_0^2 s_0(\tau_0) + \tau_1^2 s_1(\tau_1) + \tau_2^2 s_2(\tau_2) = t\, \tau_0\tau_1\tau_2. \tag{3.82}$$

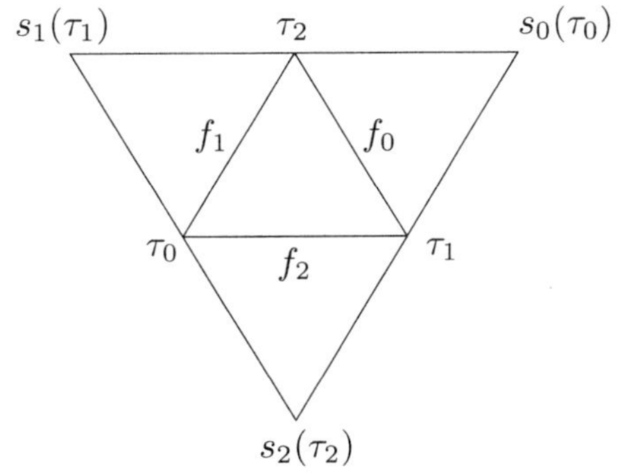

FIGURE 3.2. The Six τ-Functions

If we regard the ratio

$$[\tau_0 : \tau_1 : \tau_2 : s_0(\tau_0) : s_1(\tau_1) : s_2(\tau_2)] \in \mathbb{P}^5 \tag{3.83}$$

as a point of the 5-dimensional projective space, formula (3.82) defines a cubic hypersurface moving around in $\mathbb{P}^5$ as t varies.

3.4.2. Relations among three τ-functions on a line. Let us consider the following figure:

(3.84)

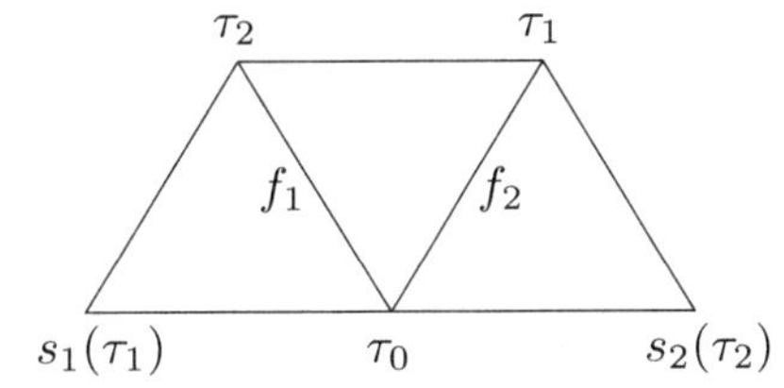

From the formulas that represent the f variables in terms of τ-functions we have

$$\tau_1 s_1(\tau_1) = f_1\, \tau_0\, \tau_2, \quad \tau_2 s_2(\tau_2) = f_2\, \tau_1\, \tau_0, \tag{3.85}$$

and hence

$$s_1(\tau_1) s_2(\tau_2) = f_1 f_2\, \tau_0^2. \tag{3.86}$$

On the other hand, from (3.39) we get

$$f_1 f_2 = h_0' - \frac{\alpha_1 - \alpha_2}{3} = \frac{\tau_0''}{\tau_0} - \left(\frac{\tau_0'}{\tau_0}\right)^2 - \frac{\alpha_1 - \alpha_2}{3}. \tag{3.87}$$

This implies

$$\begin{aligned} s_1(\tau_1)\, s_2(\tau_2) &= \tau_0'' \tau_0 - (\tau_0')^2 - \left(\frac{\alpha_1 - \alpha_2}{3}\right) \tau_0^2 \\ &= \left(\frac{1}{2} D_t^2 - \frac{\alpha_1 - \alpha_2}{3}\right) \tau_0 \cdot \tau_0. \end{aligned} \tag{3.88}$$

This relation is essentially the *Toda equation*, which we are going to discuss later. Around τ_0 we have three such equations corresponding to the three directions in the triangular lattice.

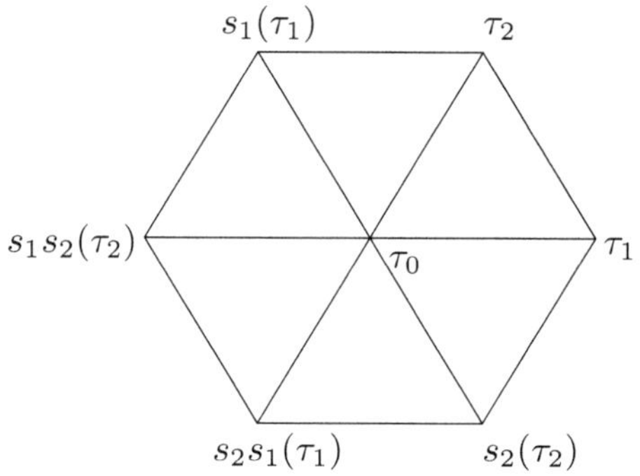

FIGURE 3.3. Around the Hexagon

3.4.3. Around the hexagon. (Figure 3.3.) We now recall a typical Bäcklund transformation for the f variables:

$$s_1(f_2) = f_2 + \frac{\alpha_1}{f_1}. \tag{3.89}$$

Let us rewrite this into the relation of τ-functions. Since

$$s_1(f_2) = s_1\left(\frac{\tau_2\, s_2(\tau_2)}{\tau_1 \tau_0}\right) = \frac{\tau_2\, s_1 s_2(\tau_2)}{s_1(\tau_1)\, \tau_0}, \tag{3.90}$$

the formula above gives rise to

$$\frac{\tau_2\, s_1 s_2(\tau_2)}{s_1(\tau_1)\, \tau_0} = \frac{\tau_2\, s_2(\tau_2)}{\tau_1 \tau_0} + \frac{\alpha_1\, \tau_0\, \tau_2}{\tau_1\, s_1(\tau_1)}, \tag{3.91}$$

and hence

$$\tau_1\, s_1 s_2(\tau_2) = s_1(\tau_1) s_2(\tau_2) + \alpha_1\, \tau_0^2. \tag{3.92}$$

Together with the one obtained similarly from $s_2(f_1)$, we have the following two relations:

$$\begin{aligned} \tau_1\, s_1 s_2(\tau_2) - s_1(\tau_1)\, s_2(\tau_2) &= \alpha_1\, \tau_0^2, \\ s_1(\tau_1)\, s_2(\tau_2) - s_2 s_1(\tau_1)\, \tau_2 &= \alpha_2\, \tau_0^2. \end{aligned} \tag{3.93}$$

Eliminating τ_0 from these two, we have a quadratic equation

$$(\alpha_0 - 1) s_1(\tau_1) s_2(\tau_2) + \alpha_1\, s_2 s_1(\tau_1) \tau_2 + \alpha_2\, \tau_1 s_1 s_2(\tau_2) = 0 \tag{3.94}$$

relating the six τ-functions placed at the vertices of a regular hexagon. These relations correspond to what are called *Hirota-Miwa equations* in the context of discrete systems.

3.5. τ-Functions for P_{II}

Also for the symmetric form of P_{II} (Section 2.5), it is possible to develop an argument of τ-functions, similar to those for the symmetric form of P_{IV}. We summarize below the results for the case of P_{II}, paying attention only to the points where differences arise. Please fill in the details by yourself when the computation is omitted.

3.5.1. The symmetric form. Recall that the symmetric form of P_{II} was obtained by rewriting P_{II} in terms of the variables

$$f_0 = -p + 2q^2 + t, \quad f_1 = p, \quad \alpha_0 = 1 - b, \quad \alpha_1 = b. \tag{3.95}$$

Explicitly it is given by

$$\begin{cases} f_0' = -2qf_0 + \alpha_0, \\ f_1' = 2qf_1 + \alpha_1, \\ q' = \dfrac{1}{2}(f_1 - f_0), \end{cases} \qquad f_0 + f_1 - 2q^2 = t. \tag{3.96}$$

The last algebraic equation is the condition for normalizing the three variables.

3.5.2. Poisson structure. The dependent variables (f_0, f_1, q) for the symmetric form are regarded as providing coordinates for the three dimensional space having the original variables (q, p, t) as coordinates. On this space we already have the usual Poisson structure defined by $\{p, q\} = 1$, $\{q, t\} = \{p, t\} = 0$. Under this Poisson structure the Poisson brackets among the variables (f_0, f_1, q) are computed as listed in the following table:

(3.97)

$\{\ ,\ \}$	f_0	f_1	q
f_0	0	$-4q$	-1
f_1	$4q$	0	1
q	1	-1	0

This table, corresponding to (3.28) in the case of P_{IV}, plays a role similar to that of the matrix U. In fact, by Table 2.2 of Bäcklund transformations for P_{II} in Section 2.5 we find that s_0 and s_1 are defined through the formulas

$$\begin{aligned} s_i(f_j) &= f_j + \frac{\alpha_i}{f_i}\{f_i, f_j\} + \frac{1}{2}\left(\frac{\alpha_i}{f_i}\right)^2 \{f_i, \{f_i, f_j\}\}, \\ s_i(q) &= q + \frac{\alpha_i}{f_i}\{f_i, q\}, \end{aligned} \tag{3.98}$$

for $i, j = 0, 1$. Let us now introduce the notation $\mathrm{ad}_{\{\}}(\varphi) = \{\varphi, \cdot\}$. Then the action of $\mathrm{ad}_{\{\}}(f_0)$ is given by

$$f_1 \to -4q \to 4 \to 0, \qquad q \to -1 \to 0, \tag{3.99}$$

so that $\mathrm{ad}_{\{\}}(f_0)^3(f_1) = 0$, $\mathrm{ad}_{\{\}}(f_0)^2(q) = 0$. (One obtains similar formulas by exchanging the roles of 0 and 1.) Accordingly the two formulas in (3.98) are unified into a single formula

$$\begin{aligned} s_i(\varphi) &= \varphi + \frac{\alpha_i}{f_i}\mathrm{ad}_{\{\}}(f_i)(\varphi) + \frac{1}{2!}\left(\frac{\alpha_i}{f_i}\right)^2 \mathrm{ad}_{\{\}}(f_i)^2(\varphi) + \cdots \\ &= \exp\left(\frac{\alpha_i}{f_i}\mathrm{ad}_{\{\}}(f_i)\right)(\varphi) \end{aligned} \tag{3.100}$$

for $\varphi = f_0, f_1, q$.

3.5.3. Hamiltonians. Using the Hamiltonian

$$H=\frac{1}{2}p^2-\left(q^2+\frac{t}{2}\right)p-bq=\frac{1}{2}p(p-q^2-t)-bq \tag{3.101}$$

for $P_{\rm II}$, we set $h_0=H$ and $h_1=\pi(H)$. (In the previous notation, $\pi=r$.) We use the two Hamiltonians

$$h_0=-\frac{1}{2}f_0f_1-\alpha_1 q,\quad h_1=-\frac{1}{2}f_0f_1+\alpha_0 q \tag{3.102}$$

for introducing the τ-functions of $P_{\rm II}$. Note that the symmetric form of $P_{\rm II}$ is expressed as

$$f_0'=\{h_0,f_0\}+1,\quad f_1'=\{h_0,f_1\},\quad q'=\{h_0,q\} \tag{3.103}$$

in terms of the Hamiltonians. From h_0 and h_1 the original variables are recovered by

$$f_0=-2h_1',\quad f_1=-2h_0',\quad q=h_1-h_0. \tag{3.104}$$

The differential equations for f_0, f_1 of the symmetric form can be represented alternatively as

$$f_0'=f_0(2h_0-2h_1)+\alpha_0,\quad f_1'=f_1(-2h_0+2h_1)+\alpha_1. \tag{3.105}$$

(You see the Cartan matrix here, too.) The Hamiltonians h_0, h_1 also carry the invariance

$$s_i(h_j)=h_j\quad(i\neq j),\quad s_j(h_j)=h_j+\frac{\alpha_j}{f_j} \tag{3.106}$$

under the actions of s_0, s_1.

3.5.4. Differential equations for the τ-functions. Let us introduce two τ-functions by

$$h_0=\frac{\tau_0'}{\tau_0},\quad h_1=\frac{\tau_1'}{\tau_1}. \tag{3.107}$$

Then the original variables are recovered by

$$\begin{aligned} f_0&=-2\left(\log\tau_1\right)''=-\frac{1}{\tau_1^2}D_t^2\,\tau_1\cdot\tau_1,\\ f_1&=-2\left(\log\tau_0\right)''=-\frac{1}{\tau_0^2}D_t^2\,\tau_0\cdot\tau_0,\\ q&=\left(\log\frac{\tau_1}{\tau_0}\right)'=\frac{1}{\tau_0\tau_1}D_t\,\tau_1\cdot\tau_0. \end{aligned} \tag{3.108}$$

Under this correspondence the original $P_{\rm II}$ is equivalent to the following system of bilinear differential equations of Hirota type:

$$\begin{aligned} &\left(D_t^2+\frac{t}{2}\right)\tau_0\cdot\tau_1=0,\\ &\left(D_t^3+\frac{t}{2}D_t-\frac{\alpha_0-\alpha_1}{2}\right)\tau_0\cdot\tau_1=0. \end{aligned} \tag{3.109}$$

The fundamental rational solution

$$(\alpha_0,\alpha_1;f_0,f_1,q)=\left(\frac{1}{2},\frac{1}{2};\frac{t}{2},\frac{t}{2},0\right) \tag{3.110}$$

gives rise to the τ-functions

$$(\tau_0,\tau_1)=\left(c_0\exp\left(-\frac{t^3}{24}\right),c_1\exp\left(-\frac{t^3}{24}\right)\right). \tag{3.111}$$

The Riccati solution at $\alpha_1=0$

$$(\alpha_0,\alpha_1;f_0,f_1,q)=\left(1,0;2\left(\frac{\varphi'}{\varphi}\right)^2+t,0,\frac{\varphi'}{\varphi}\right);\quad \varphi''+\frac{t}{2}\varphi=0 \tag{3.112}$$

provides the τ-functions

$$(\tau_0,\tau_1)=(c_0,c_1\varphi). \tag{3.113}$$

3.5.5. Bäcklund transformations of the τ-functions. The Bäcklund transformations of the τ-functions can be defined by

$$\begin{aligned} s_0(\tau_0)&=f_0\frac{\tau_1^2}{\tau_0}=-\frac{1}{\tau_0}D_t^2\,\tau_1\cdot\tau_1,\\ s_1(\tau_1)&=f_1\frac{\tau_0^2}{\tau_1}=-\frac{1}{\tau_1}D_t^2\,\tau_0\cdot\tau_0,\end{aligned} \tag{3.114}$$

together with the invariance $s_i(\tau_j)=\tau_j$ $(i\neq j)$. (The Bäcklund transformation $\pi=r$ exchanges τ_0 and τ_1 with each other.) These also imply the multiplicative formulas

$$f_0=\frac{\tau_0\,s_0(\tau_0)}{\tau_1^2},\qquad f_1=\frac{\tau_1\,s_1(\tau_1)}{\tau_0^2} \tag{3.115}$$

for f_j.

3.5.6. Relations among τ-functions. Let us consider the sequence of τ-functions

$$\begin{array}{cccccc} \tau_{-2} & \tau_{-1} & \tau_0 & \tau_1 & \tau_2 & \tau_3\\ \| & \| & & & \| & \|\\ s_1s_0(\tau_0) & s_1(\tau_1) & & & s_0(\tau_0) & s_0s_1(\tau_1)\end{array} \tag{3.116}$$

obtained as Bäcklund transformations of τ_0, τ_1. From the normalization condition $f_0+f_1-2q^2=t$, we have

$$\frac{\tau_0\tau_2}{\tau_1^2}+\frac{\tau_{-1}\tau_1}{\tau_0^2}-2q^2=t,\quad q=\frac{\tau_0\tau_1'-\tau_0'\tau_1}{\tau_0\tau_1}. \tag{3.117}$$

Also from the formulas for the Bäcklund transformations $s_0(f_1)$, $s_1(f_0)$ we have

$$\begin{aligned} &\frac{\tau_{-2}\tau_1}{\tau_{-1}\tau_0}-\frac{\tau_0\tau_3}{\tau_1\tau_2}+\frac{2\alpha_0^2\,\tau_1^2}{\tau_0\tau_2}-\frac{2\alpha_1^2\,\tau_0^2}{\tau_{-1}\tau_1}=4q,\\ &\frac{\tau_{-1}\tau_2}{\tau_0\tau_1}-\frac{\alpha_0\,\tau_{-2}\tau_1}{\tau_{-1}\tau_0}-\frac{\alpha_1\,\tau_0\tau_3}{\tau_1\tau_2}+\frac{2\alpha_0^2\alpha_1\,\tau_1^2}{\tau_0\tau_2}+\frac{2\alpha_0\alpha_1^2\,\tau_0^2}{\tau_{-1}\tau_1}=0.\end{aligned} \tag{3.118}$$

(We omit the detail of computation for deriving these equations.) Eliminate q by combining (3.117) and (3.118) in order to obtain two algebraic relations among the six τ-functions $\tau_{-2},\dots,\tau_3$.

CHAPTER 4

τ-Functions on the Lattice

In this chapter we formulate the τ-functions on the lattice in order to provide a method to capture the whole set of solutions obtained by Bäcklund transformations. We apply this method to analyze the structure of special polynomials such as Yablonski-Vorobiev polynomials and Okamoto polynomials, associated with the rational solutions of Painlevé equations. In the latter half of this chapter we extend the framework of τ-functions on the lattice to those of type $A_l^{(1)}$ from the viewpoint of discrete systems.

4.1. The lattice for $P_{\rm II}$

Recall that the symmetric form of $P_{\rm II}$, as discussed in Sections 2.5 and 3.5, was derived by rewriting the Hamiltonian system $H_{\rm II}$ of Proposition 1.1 by means of the dependent variables $f_0 = -p+2q^2+t$, $f_1 = p$, q and the parameters $\alpha_0 = 1-b$, $\alpha_1 = b$. In terms of these variables we described the action of the extended affine Weyl group $\widetilde{W} = \langle\, s_0, s_1, \pi\,\rangle$, and introduced the τ-functions τ_0 and τ_1 for $P_{\rm II}$.

We now define the transformation T by setting $T = \pi s_1 = rs$. Then we have

$$
\begin{aligned}
&T(\alpha_0) = \alpha_0 + 1, \quad T(\alpha_1) = \alpha_1 - 1,\\
&T(f_0) = f_1 - \frac{4\alpha_0 q}{f_0} + \frac{2\alpha_0^2}{f_0^2}, \quad T(f_1) = f_0, \quad T(q) = -q + \frac{\alpha_0}{f_0},\\
&T(\tau_0) = \tau_1, \quad T(\tau_1) = f_0\,\frac{\tau_1^2}{\tau_0}.
\end{aligned}
\tag{4.1}
$$

For a general solution with parameters (α_0, α_1) the rational functions above represent a solution with parameters (α_0+1, α_1-1). Namely, they provide the explicit formulas for expressing the new solution resulting from the transformation in terms of the original dependent variables. For a general integer $n \in \mathbb{Z}$ one can also consider the transformation T^n obtained by the iteration of T. Since

$$
T^n(\alpha_0) = \alpha_0 + n, \quad T^n(\alpha_1) = \alpha_1 - n \qquad (n \in \mathbb{Z}),
\tag{4.2}
$$

T^n is understood as the Bäcklund transformation that produces a solution with parameter $\alpha_1 - n = b - n$ from a general solution with parameter $\alpha_1 = b$. We discuss below the problem of how the transforms $T^n(f_j)$, $T^n(q)$, $T^n(\tau_j)$ can be expressed explicitly as functions of the original f_j, q, τ_j. For that purpose we begin by considering the structure of transformations of the τ-functions.

Noting that $T(\tau_0) = \tau_1$, we introduce the notation

$$
\tau_n = T^n(\tau_0) \qquad (n \in \mathbb{Z})
\tag{4.3}
$$

for the transforms of the τ-functions. In this way we obtain a sequence of τ-functions

$$\cdots \xrightarrow{T} \tau_{n-1} \xrightarrow{T} \tau_n \xrightarrow{T} \tau_{n+1} \xrightarrow{T} \cdots \tag{4.4}$$

indexed by the integers. We remark that in this infinite sequence each consecutive pair $(\tau_n, \tau_{n+1}) = (T^n(\tau_0), T^n(\tau_1))$ forms the pair of τ-functions for the solution at $\alpha_1 - n$.

PROPOSITION 4.1. *The action of* $\widetilde{W} = \langle s_0, s_1, \pi \rangle$ *on the* τ*-functions* τ_n $(n \in \mathbb{Z})$ *is given by*

$$s_0(\tau_n) = \tau_{2-n}, \quad s_1(\tau_n) = \tau_{-n}, \quad \pi(\tau_n) = \tau_{1-n} \quad (n \in \mathbb{Z}). \tag{4.5}$$

The proof will be left to the reader as an exercise. We only remark that as for the index n for the τ-functions the actions of s_0, s_1 and π represent the mirror-image (reflection) at $n = 1$, $n = 0$ and $n = \frac{1}{2}$ respectively.

By applying T^n to the relations (3.109), (3.114) presented in Section 3.5, we obtain the following equations for the sequence of τ-functions τ_n.

THEOREM 4.2. *The sequence of* τ*-functions* τ_n $(n \in \mathbb{Z})$ *satisfies the following bilinear equations.*

(1) *Bilinear differential equations of Hirota type*:

$$\begin{aligned} &\left(D_t^2 + \frac{t}{2}\right) \tau_n \cdot \tau_{n+1} = 0, \\ &\left(D_t^3 + \frac{t}{2} D_t - \frac{\alpha_0 - \alpha_1}{2} - n\right) \tau_n \cdot \tau_{n+1} = 0. \end{aligned} \tag{4.6}$$

(2) *Toda equation*:

$$\tau_{n-1}\tau_{n+1} = -D_t^2\, \tau_n \cdot \tau_n = -2\left(\tau_n''\tau_n - (\tau_n')^2\right). \tag{4.7}$$

Let us make some experiments to see how the general τ-functions τ_n are expressed in terms of the initial variables τ_0, τ_1:

$$\tau_2 = f_0\, \frac{\tau_1^2}{\tau_0}, \quad \tau_3 = r s_1(f_0)\, f_0^2 \frac{\tau_1^3}{\tau_0^2} = (f_0^2 f_1 - 4\alpha_0 f_0 q + 2\alpha_0^2)\frac{\tau_1^3}{\tau_0^2}, \dots. \tag{4.8}$$

For general $n \in \mathbb{Z}$, one can show inductively that there exists a rational function ϕ_n of f_0, f_1, q, α_0, α_1 such that τ_n is expressed in the form

$$\tau_n = T^n(\tau_0) = \phi_n\, \frac{\tau_1^n}{\tau_0^{n-1}} \quad (n \in \mathbb{Z}). \tag{4.9}$$

Here ϕ_n are determined by the recurrence formulas

$$\phi_{n+1} = T(\phi_n)\, f_0^n, \quad \phi_{n-1} = T^{-1}(\phi_n)\, f_1^{1-n} \quad (n \in \mathbb{Z}) \tag{4.10}$$

with initial value $\phi_0 = 1$. Some ϕ_n with small n are determined explicitly as

$$\begin{aligned} &\phi_{-2} = f_0 f_1^2 + 4\alpha_1 f_1 q + 2\alpha_1^2, \quad \phi_{-1} = f_1, \quad \phi_0 = 1, \\ &\phi_1 = 1, \quad \phi_2 = f_0, \quad \phi_3 = f_0^2 f_1 - 4\alpha_0 f_0 q + 2\alpha_0^2. \end{aligned} \tag{4.11}$$

In what follows we will simply call the ϕ_n the *ϕ-factors*. (This is a term used only in this book.) By definition these ϕ_n must be rational functions, but in view of the examples above it is expected that they would be in fact polynomials. Let us consider the case of the rational solution

$$(\alpha_0, \alpha_1; f_0, f_1, q) = \left(\frac{1}{2}, \frac{1}{2}; \frac{t}{2}, \frac{t}{2}, 0\right). \tag{4.12}$$

In this case, starting from $\tau_0 = \tau_1 = \exp(-\frac{t^3}{24})$ we obtain

$$T^n(\tau_0) = \phi_n \exp\left(-\frac{t^3}{24}\right), \quad T^n(\tau_1) = \phi_{n+1} \exp\left(-\frac{t^3}{24}\right), \tag{4.13}$$

which give the two τ-functions for the rational solution at $b = \frac{1}{2} - n$ in the original parametrization. The ϕ-factors for this case are computed by specialization as follows:

$$\begin{aligned} &\phi_{-2} = \frac{1}{8}(t^3+4), \quad \phi_{-1} = \frac{t}{2}, \quad \phi_0 = 1, \\ &\phi_1 = 1, \quad \phi_2 = \frac{t}{2}, \quad \phi_3 = \frac{1}{8}(t^3+4). \end{aligned} \tag{4.14}$$

Compare these polynomials with the Yablonski-Vorobiev polynomials P_n which we have already seen in Section 1.5:

$$P_0 = P_1 = 1, \quad P_2 = P_{-1} = t, \quad P_3 = P_{-2} = t^3 + 4. \tag{4.15}$$

The ϕ-factors, which appear in the expression of τ_n in terms of the initial data τ_0, τ_1, can be regarded as the counterpart of the Yablonski-Vorobiev polynomials for the general solutions. Here we propose two ways of expressing $T^n(f_j)$ by using ϕ_n. Recall first the formula for determining f_1 from the τ-functions:

$$f_1 = \frac{\tau_1\, s_1(\tau_1)}{\tau_0^2} = \frac{\tau_{-1}\,\tau_1}{\tau_0^2}. \tag{4.16}$$

By applying T^n to this formula, $T^n(f_1)$ is computed as

$$T^n(f_1) = \frac{\tau_{n-1}\,\tau_{n+1}}{\tau_n^2} = \frac{\phi_{n-1}\,\phi_{n+1}}{\phi_n^2}\frac{\tau_0^{2-n}\tau_1^{n-1}\;\tau_0^{-n}\tau_1^{n+1}}{(\tau_0^{1-n}\tau_1^n)^2} = \frac{\phi_{n-1}\,\phi_{n+1}}{\phi_n^2}, \tag{4.17}$$

where one can see that the factors of τ_0, τ_1 appearing in the numerator and the denominator nicely cancel out. (This is not surprising, since we know from the very beginning that $T^n(f_j)$ should be a rational function of f_j, q and α_j.) Also from the formula

$$f_0 = -2(\log\tau_1)'', \quad f_1 = -2(\log\tau_0)'', \tag{4.18}$$

representing f_j in terms of the derivatives of τ_j, we obtain

$$\begin{aligned} T^n(f_1) &= -2(\log\tau_n)'' = -2(\log\phi_n\tau_1^n\tau_0^{1-n})'' \\ &= nf_0 - (n-1)f_1 - 2(\log\phi_n)''. \end{aligned} \tag{4.19}$$

Note that since the two formulas (4.17) and (4.19) should give the same answer, we must have

$$\frac{\phi_{n-1}\phi_{n+1}}{\phi_n^2} = nf_0 - (n-1)f_1 - 2(\log\phi_n)''. \tag{4.20}$$

We remark that this formula is nothing but the Toda equation for the ϕ-factors, which in fact follows from (4.7); in this sense the two expressions (4.17) and (4.19) are consistent with each other. From this remark we also conclude that ϕ_n can be determined by the recurrence formula

$$\phi_{n-1}\phi_{n+1} = -2\left(\phi_n''\phi_n - (\phi_n')^2\right) + (nf_0 - (n-1)f_1)\phi_n^2 \tag{4.21}$$

with the initial condition $\phi_0 = \phi_1 = 1$.

Combining these arguments with similar computations for f_0 and q, we obtain

THEOREM 4.3. *For each $n \in \mathbb{Z}$, $\tau_n = T^n(\tau_0)$ is expressed as $\tau_n = \phi_n\, \tau_1^n/\tau_0^{n-1}$, where ϕ_n is the rational function in f_j, q, α_j determined by the recurrence formula (4.10) (or the Toda equation (4.21)). With these ϕ-factors we have*

$$\begin{aligned} T^n(f_0) &= \frac{\phi_n\,\phi_{n+2}}{\phi_{n+1}^2} = (n+1)f_0 - nf_1 - 2(\log\phi_{n+1})'', \\ T^n(f_1) &= \frac{\phi_{n-1}\,\phi_{n+1}}{\phi_n^2} = nf_0 - (n-1)f_1 - 2(\log\phi_n)'', \\ T^n(q) &= q + \frac{\phi_{n+1}'\phi_n - \phi_{n+1}\phi_n'}{\phi_n\phi_{n+1}} = q + \left(\log\frac{\phi_{n+1}}{\phi_n}\right)' \end{aligned} \tag{4.22}$$

for each $n \in \mathbb{Z}$.

As for the ϕ-factors for the τ-functions, it is known that:

Each ϕ_n ($n \in \mathbb{Z}$) is a polynomial in α_j, f_j ($j = 0, 1$) and q.

We will discuss in later chapters how this can be the case, and we continue on by assuming this statement for the present. If by any means we know that they are polynomials, we can prove the following lemma by induction.

LEMMA 4.4. *Each ϕ_n ($n \in \mathbb{Z}$) is a polynomial in f_0, f_1 of degree $n(n-1)/2$. Its leading term is determined as follows:*

$$\begin{aligned} &f_0^{m^2} f_1^{m(m-1)} && (\text{when } n \text{ is even and } n = 2m), \\ &f_0^{m(m+1)} f_1^{m^2} && (\text{when } n \text{ is odd and } n = 2m+1). \end{aligned} \tag{4.23}$$

Substituting the rational solution (4.12), we see that each ϕ_n is expressed as

$$\phi_n = \frac{1}{2^{n(n-1)/2}} P_n(t), \quad P_n(t) = t^{n(n-1)/2} + \text{lower degree terms} \tag{4.24}$$

with a monic polynomial $P_n = P_n(t)$ of degree $n(n-1)/2$. Hence we obtain the following expressions for $T^n(f_j)$ and $T^n(q)$:

$$\begin{aligned} &T^n(f_0) = \frac{P_nP_{n+2}}{2P_{n+1}^2}, \quad T^n(f_1) = T^n(p) = \frac{P_{n-1}P_{n+1}}{2P_n^2}, \\ &T^n(q) = \frac{P_{n+1}'P_n - P_{n+1}P_n'}{P_nP_{n+1}}. \end{aligned} \tag{4.25}$$

The polynomials P_n appearing in these expressions are the *Yablonski-Vorobiev polynomials.*

Using the example of $P_{\rm II}$, at this point we make a remark about how one can regard the action of (4.1) of T as a system of difference equations. Since $f_0 + f_1 - 2q^2 = t$, we can eliminate f_0 in (4.1) to have

$$T(f_1) = t - f_1 + 2q^2, \quad T(q) = -q + \frac{\alpha_0}{t - f_1 + 2q^2}. \tag{4.26}$$

If we apply T^n to these equations, the resulting formulas give rise to the system of difference equations

$$x_n + x_{n+1} = 2y_n^2 + t, \quad y_{n-1} + y_n = \frac{\alpha_0 + n - 1}{x_n} \tag{4.27}$$

for $x_n = T^n(f_1)$ and $y_n = T^n(q)$. Here we regard n as the independent variable, and t, α_0 as parameters. Theorem 4.3 mentioned above is then understood as providing a method for constructing a general solution

$$x_n = \frac{\phi_{n-1}\phi_{n+1}}{\phi_n^2}, \quad y_n = y_0 + \frac{\phi'_{n+1}\phi_n - \phi_{n+1}\phi'_n}{\phi_n\phi_{n+1}} \tag{4.28}$$

of the difference system (4.27) by means of differentiation with respect to the parameter t.

4.2. The lattice for $P_{\rm IV}$

We now consider a similar construction for the symmetric form of $P_{\rm IV}$. In the parameter space of this case there are several possible directions of parallel translations that preserve the triangular lattice. We first explain a general idea concerning a method of describing the parallel translations by means of the extended affine Weyl group. After that, we will formulate the τ-functions on the lattice for the case of $P_{\rm IV}$.

We consider the triangular lattice as in the figure below and denote by C the fundamental regular triangle $\Delta P_0 P_1 P_2$. We denote by T_1 the parallel translation specified by the vector $\overrightarrow{P_0P_1}$.

(4.29)

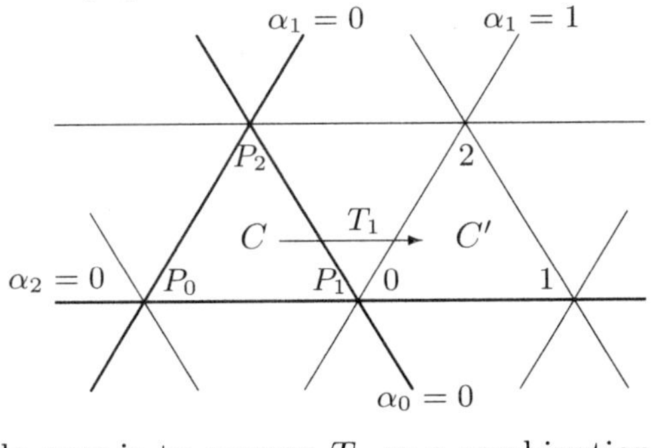

What we want to do now is to express T_1 as a combination of s_0, s_1, s_2, π. We first look at the triangle $C' = T_1(C)$ obtained by transferring the fundamental triangle C by T_1. Then we try to move C' back to the location of C by using only s_0, s_1, s_2. Notice that there are three lines (reflection lines) $\alpha_0 = 0$, $\alpha_1 = 0$, $\alpha_2 = 0$ surrounding the fundamental triangle C. Among these we choose one line separating (sitting in between) C and C'. Since the line $\alpha_0 = 0$ is the only such choice for now, we turn over C' by using s_0. Then as $\alpha_2 = 0$ is the only line that separates $s_0(C')$ and C, we use s_2 to turn over $s_0(C')$. The triangle $s_2s_0(C')$ thus obtained overlaps with C so that $s_2s_0(C') = C$, i.e. $s_2s_0T(C) = C$. Although

C' has come back to the original position, $s_2 s_0 T$ still differs from the identify transformation. This is because the three points P_0, P_1, P_2 have not yet come back to their right positions. We need further to rotate $s_2 s_0(C')$ by $-120°$ to finish our job. The composition $\pi^{-1} s_2 s_0 T$ then reduces to the identity transformation, since the identity transformation is the only affine transformation of the plane that leaves three distinct points fixed. After all we have $\pi^{-1} s_2 s_0 T = 1$, namely, $T = s_0 s_2 \pi = \pi s_2 s_1$.

(4.30)

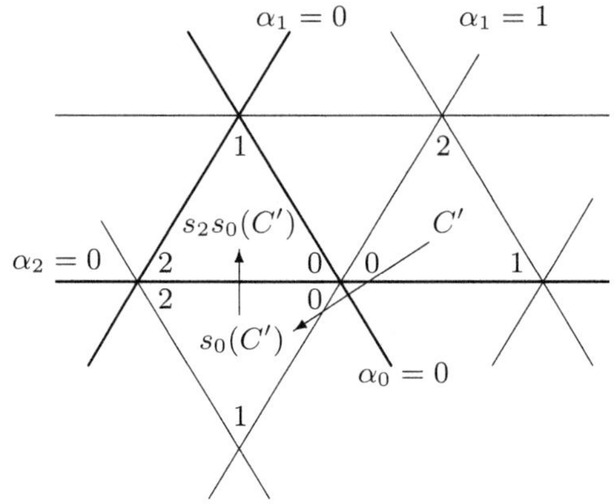

Remark: By the procedure described above we see that the fundamental triangle C can be transferred to an arbitrary triangle C' by some element $w \in W = \langle s_0, s_1, s_2 \rangle$. For any choice of C' there are only a finite number of reflection lines $\alpha_j = n$ ($j = 0, 1, 2$; $n \in \mathbb{Z}$) that separate C and C'. If we repeat the procedure above, at each step the number of lines lying in between decreases by one; the triangle C' will finally be laid over C when the number becomes zero. Then, by reversing the procedure we obtain an element $w \in W$ such that $w(C) = C'$. This shows that C is a *fundamental domain* under the action of $W = \langle s_0, s_1, s_2 \rangle$ on the plane. However one cannot produce all the translations unless one uses such transformations that map C to itself, corresponding to automorphisms of the Dynkin diagram. This is the reason why we consider the extended affine Weyl group $\widetilde{W} = \langle s_0, s_1, s_2, \pi \rangle$.

From the discussion above, it follows that the parallel translations with respect to the three vectors $\overrightarrow{P_0P_1}$, $\overrightarrow{P_1P_2}$, $\overrightarrow{P_2P_0}$ are realized as elements of $\widetilde{W} = \langle s_0, s_1, s_2, \pi \rangle$, and expressed as

$$T_1 = \pi s_2 s_1, \quad T_2 = \pi T_1 \pi^{-1} = s_1 \pi s_2, \quad T_3 = \pi T_2 \pi^{-1} = s_2 s_1 \pi, \tag{4.31}$$

respectively. These three parallel translations commute with each other, so that

$$T_i T_j = T_j T_i \quad (i, j = 1, 2, 3); \quad T_1 T_2 T_3 = 1. \tag{4.32}$$

We remark that this fact can also be proved directly from the fundamental relations for the generators of $\widetilde{W}$.

Exercise: Regarding (4.31) as the definition of T_i ($i = 1, 2, 3$), prove the relations (4.32) by using only the fundamental relations of $\widetilde{W}$.

We now regard these T_1, T_2, T_3 as the Bäcklund transformations of the symmetric form of P_{IV}:

$$T_1 = \pi s_2 s_1 : \quad \alpha_0 \xrightarrow{s_1} \alpha_0 + \alpha_1 = -\alpha_2 + 1 \xrightarrow{s_2} \alpha_2 + 1 \xrightarrow{\pi} \alpha_0 + 1. \tag{4.33}$$

Hence we have $T_1(\alpha_0) = \alpha_0 + 1$. Similarly, we can compute the action of T_i on α_j $(j = 0, 1, 2)$ as follows:

$$
\begin{aligned}
&T_1(\alpha_0) = \alpha_0 + 1, \quad T_1(\alpha_1) = \alpha_1 - 1, \quad T_1(\alpha_2) = \alpha_2 \\
&T_2(\alpha_0) = \alpha_0, \quad T_2(\alpha_1) = \alpha_1 + 1, \quad T_2(\alpha_2) = \alpha_2 - 1 \\
&T_3(\alpha_0) = \alpha_0 - 1, \quad T_3(\alpha_1) = \alpha_1, \quad T_3(\alpha_2) = \alpha_2 + 1.
\end{aligned}
\tag{4.34}
$$

This computation implies for instance that the Bäcklund transformation T_1, applied to a solution with parameters $(\alpha_0, \alpha_1, \alpha_2)$, gives rise to a solution with parameters $(\alpha_0 + 1, \alpha_1 - 1, \alpha_2)$. (Notice that the points in the parameter space have moved in the direction opposite to $\overrightarrow{P_0P_1}$. The directions in which the points and the *observer* move are opposite to each other. One can also think the action of T_1 on α_j as corresponding to moving the axes for the triangular coordinates in the direction of $\overrightarrow{P_0P_1}$.) The other two translations T_2 and T_3 can be understood similarly. Let us now examine the action of $T_1 = \pi s_2 s_1$ on f_j and τ_j:

$$
f_0 \overset{s_1}{\to} f_0 - \frac{\alpha_1}{f_1} \overset{s_2}{\to} f_0 + \frac{\alpha_2}{f_2} - \frac{\alpha_1 + \alpha_2}{f_1 - \frac{\alpha_2}{f_2}}.
\tag{4.35}
$$

Hence we have

$$
\begin{aligned}
T_1(f_0) &= f_1 + \frac{\alpha_0}{f_0} - \frac{\alpha_2 + \alpha_0}{f_1 - \frac{\alpha_0}{f_0}} \\
&= \frac{f_0^2 f_1 f_2 - \alpha_0 f_0 f_1 + \alpha_0 f_2 f_0 - (\alpha_0 + \alpha_2) f_0^2 - \alpha_0^2}{f_0(f_0 f_2 - \alpha_0)}.
\end{aligned}
\tag{4.36}
$$

This example shows how the combination of the transformations s_i produces continued fractions, and also what happens when the continued fractions are factored. Similarly we have

$$
\begin{aligned}
T_1(f_1) &= f_2 - \frac{\alpha_0}{f_0} = \frac{f_0 f_2 - \alpha_0}{f_0}, \\
T_1(f_2) &= f_0 + \frac{\alpha_0 + \alpha_2}{f_2 - \frac{\alpha_0}{f_0}} = \frac{f_0(f_0 f_2 + \alpha_2)}{f_0 f_2 - \alpha_0}.
\end{aligned}
\tag{4.37}
$$

The τ-functions in turn transform as follows:

$$
T_1(\tau_0) = \tau_1, \quad T_1(\tau_1) = (f_0 f_2 - \alpha_0)\frac{\tau_1^2}{\tau_0}, \quad T_1(\tau_2) = f_0 \frac{\tau_1 \tau_2}{\tau_0}.
\tag{4.38}
$$

I recommend the reader to do experiments to see what happens when one performs the Bäcklund transformations by combining T_1, T_2, T_3 repeatedly.

In order to describe the Bäcklund transformations by T_1, T_2, T_3 in a reasonable way, it is convenient to introduce the family of τ-functions

$$
\tau_{l,m,n} = T_1^l T_2^m T_3^n(\tau_0) \qquad (l, m, n \in \mathbb{Z}),
\tag{4.39}
$$

regarding τ_0 as the initial datum. Since $T_1 T_2 T_3 = 1$ we have

$$
\tau_{l+k,m+k,n+k} = \tau_{l,m,n} \qquad (k \in \mathbb{Z}).
\tag{4.40}
$$

When we want to eliminate this redundancy, we will choose the two independent directions T_1, T_2 so that

$$
\tau_{m,n} = \tau_{m,n,0} = T_1^m T_2^n(\tau_0).
\tag{4.41}
$$

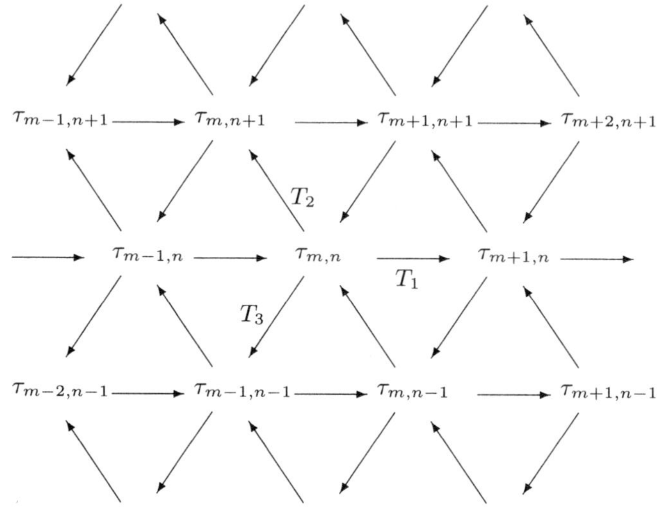

FIGURE 4.1. τ-Functions on the Lattice: $\tau_{m,n} = \tau_{m,n,0}$

In Figure 4.1 we have arranged the family of τ-functions on the triangular lattice in the plane, taking $\tau_{m,n} = \tau_{m,n,0}$ as representatives.

LEMMA 4.5. (1) *With the notation as above,*

$$\tau_0 = \tau_{0,0}, \quad \tau_1 = \tau_{1,0}, \quad \tau_2 = \tau_{1,1}. \tag{4.42}$$

(2) *The actions of* s_0, s_1, s_2, π *on* $\tau_{l,m,n}$ *are given by the formulas*

$$\begin{aligned} &s_0(\tau_{l,m,n}) = \tau_{n+1,m,l-1}, \quad s_1(\tau_{l,m,n}) = \tau_{m,l,n}, \quad s_2(\tau_{l,m,n}) = \tau_{l,n,m}, \\ &\pi(\tau_{l,m,n}) = \tau_{n+1,l,m}. \end{aligned} \tag{4.43}$$

(3) *The actions of* s_0, s_1, s_2, π *on* $\tau_{m,n}$ *are given by the formulas*

$$\begin{aligned} &s_0(\tau_{m,n}) = \tau_{2-m,1-m+n}, \quad s_1(\tau_{m,n}) = \tau_{n,m}, \quad s_2(\tau_{m,n}) = \tau_{m-n,-n}, \\ &\pi(\tau_{m,n}) = \tau_{1-n,m-n}. \end{aligned} \tag{4.44}$$

We omit the proof of this lemma since it can be verified directly by using the definition. This lemma guarantees the consistency between the Bäcklund transformations of the τ-functions and the transformations of the discrete variables of indices.

The various relations among τ-functions discussed in Chapter 3 are transformed into those among the τ-functions on the lattice simply by applying $T_1^l T_2^m T_m^n$. In fact, from Theorem 3.5 we obtain the bilinear equations of Hirota type for adjacent

τ-functions:

$$
\begin{aligned}
&\left(D_t^2+\frac{t}{3}D_t-\frac{2t^2}{9}+\frac{\alpha_0-\alpha_1+2l-m-n}{3}\right)\tau_{l,m,n}\cdot\tau_{l+1,m,n}=0,\\
&\left(D_t^2+\frac{t}{3}D_t-\frac{2t^2}{9}+\frac{\alpha_1-\alpha_2-l+2m-n+1}{3}\right)\tau_{l,m,n}\cdot\tau_{l,m+1,n}=0,\\
&\left(D_t^2+\frac{t}{3}D_t-\frac{2t^2}{9}+\frac{\alpha_2-\alpha_0-l-m+2n+2}{3}\right)\tau_{l,m,n}\cdot\tau_{l,m,n+1}=0.
\end{aligned}
\tag{4.45}
$$

(These formulas are not completely symmetric with respect to l, m, n since we have chosen τ_0 as the initial point.) On the other hand, from the relations (3.88) for three τ-functions on a line we obtain the *Toda equations* in the three directions. We will give only the formula for the T_1 direction:

$$
\tau_{l-1,m,n}\tau_{l+1,m,n}=\left(\frac{1}{2}D_t^2-\frac{\alpha_0-\alpha_1+2l-m-n-1}{3}\right)\tau_{l,m,n}\cdot\tau_{l,m,n}.
\tag{4.46}
$$

Applying $T_1^lT_2^mT_3^n$ to (3.93), we have

$$
\tau_{l-1,m,n}\tau_{l+1,m,n}-\tau_{l,m-1,n}\tau_{l,m+1,n}=(\alpha_1-l+m)\tau_{l,m,n}^2.
\tag{4.47}
$$

(This formula can also be obtained as the difference of the Toda equations in the two directions T_1 and T_2.) Similarly the relation corresponding to (3.94) is given by

$$
\begin{aligned}
&(\alpha_0-n+l-1)\,\tau_{l,m-1,n}\,\tau_{l,m+1,n}\\
&\quad+(\alpha_1-l+m)\,\tau_{l,m,n-1}\,\tau_{l,m,n+1}\\
&\quad+(\alpha_2-m+n)\,\tau_{l-1,m,n}\,\tau_{l+1,m,n}=0.
\end{aligned}
\tag{4.48}
$$

A difference equation of this form is sometimes called a *Hirota-Miwa equation.*

The action of T_i on the variables f_j can also be regarded as a difference equation (discrete dynamical system) on the triangular lattice. We now look at the T_1-direction and set

$$
x_n=T_1^n(f_1),\quad y_n=T_1^n(f_0)\quad (n\in\mathbb{Z}).
\tag{4.49}
$$

Then from

$$
\begin{aligned}
T_1(f_1)&=f_2-\frac{\alpha_0}{f_0}=t-f_0-f_1-\frac{\alpha_0}{f_0},\\
T_1^{-1}(f_0)&=f_2+\frac{\alpha_1}{f_1}=t-f_0-f_1+\frac{\alpha_1}{f_1},
\end{aligned}
\tag{4.50}
$$

we have the difference system

$$
\begin{aligned}
x_n+x_{n+1}&=t-y_n-\frac{\alpha_0+n}{y_n},\\
y_{n-1}+y_n&=t-x_n+\frac{\alpha_1-n}{x_n}.
\end{aligned}
\tag{4.51}
$$

Here n is the independent variable and t, α_0, α_1 are the parameters. We remark that this equation (4.51) is a version of the *discrete Painlevé equation* $dP_{\rm II}$; it passes to the Painlevé equation $P_{\rm II}$ by an appropriate continuous limit. (If we set $r_0=s_0s_1s_0$ and $r_1=s_2$, the affine Weyl group $\langle r_0,r_1\rangle$ of type $A_1^{(1)}$ acts on this system of difference equations. This action passes to the continuous limit and gives rise to Bäcklund transformations for $P_{\rm II}$.) The difference system for the f variables on the triangular lattice, defined as a combination of three $dP_{\rm II}$'s in the directions

T_1, T_2, T_3, is a discrete dynamical system *equivariant* under the action of the affine Weyl group $W = \langle s_0, s_1, s_2 \rangle$.

4.3. ϕ-Factors and Okamoto polynomials

Let us compare the τ-functions $\tau_{l,m,n}$ $(l, m, n \in \mathbb{Z})$ on the lattice, introduced in the previous section, with the original triple of τ-functions τ_0, τ_1, τ_2.

LEMMA 4.6. *For any $l, m, n \in \mathbb{Z}$, $\tau_{l,m,n}$ is expressed in the form*

$$\tau_{l,m,n} = \phi_{l,m,n}\, \tau_0 \left(\frac{\tau_1}{\tau_0}\right)^l \left(\frac{\tau_2}{\tau_1}\right)^m \left(\frac{\tau_0}{\tau_2}\right)^n = \phi_{l,m,n}\, \tau_0^{1+n-l}\, \tau_1^{l-m}\, \tau_2^{m-n}. \tag{4.52}$$

Here $\phi_{l,m,n}$ is a rational function in α_j, f_j determined by the recurrence relations

$$\begin{aligned} &\phi_{0,0,0} = 1, \\ &\phi_{l+1,m,n} = T_1(\phi_{l,m,n})\, f_0^{m-n}\, (f_0 f_2 - \alpha_0)^{l-m}, \\ &\phi_{l,m+1,n} = T_1(\phi_{l,m,n})\, f_1^{1+n-l}\, (f_1 f_0 - \alpha_1)^{m-n}, \\ &\phi_{l,m,n+1} = T_1(\phi_{l,m,n})\, f_2^{l-m}\, (f_2 f_1 - \alpha_2)^{1+n-l}. \end{aligned} \tag{4.53}$$

The proof can be done inductively. The recurrence formulas for $\phi_{l,m,n}$ are in fact obtained from the action of T_i on τ_j as in (4.38). For $\phi_{l,m,n}$ we have

$$\phi_{l+k,m+k,n+k} = \phi_{l,m,n} \qquad (k \in \mathbb{Z}). \tag{4.54}$$

We will use the abbreviation $\phi_{m,n} = \phi_{m,n,0}$.

Recall now the formula for representing the f variables in terms of the τ-functions. The multiplicative formula (3.80) can be written as

$$f_0 = \frac{\tau_{0,0}\tau_{2,1}}{\tau_{1,0}\tau_{1,1}}, \quad f_1 = \frac{\tau_{1,0}\tau_{0,1}}{\tau_{0,0}\tau_{1,1}}, \quad f_2 = \frac{\tau_{1,1}\tau_{0,-1}}{\tau_{0,0}\tau_{1,0}}, \tag{4.55}$$

by using $\tau_{m,n} = \tau_{m,n,0}$. Applying $T_1^m T_2^n$ to these formulas, we obtain the expressions

$$\begin{aligned} T_1^m T_2^n(f_0) &= \frac{\tau_{m,n}\tau_{m+2,n+1}}{\tau_{m+1,n}\tau_{m+1,n+1}} = \frac{\phi_{m,n}\phi_{m+2,n+1}}{\phi_{m+1,n}\phi_{m+1,n+1}}, \\ T_1^m T_2^n(f_1) &= \frac{\tau_{m+1,n}\tau_{m,n+1}}{\tau_{m,n}\tau_{m+1,n+1}} = \frac{\phi_{m+1,n}\phi_{m,n+1}}{\phi_{m,n}\phi_{m+1,n+1}}, \\ T_1^m T_2^n(f_2) &= \frac{\tau_{m+1,n+1}\tau_{m,n-1}}{\tau_{m,n}\tau_{m+1,n}} = \frac{\phi_{m+1,n+1}\phi_{m,n-1}}{\phi_{m,n}\phi_{m+1,n}}. \end{aligned} \tag{4.56}$$

When we rewrite $\tau_{m,n}$ as the product of $\phi_{m,n}$ and a monomial of τ_0, τ_1, τ_2, the factors including τ_i cancel out as before between the numerator and the denominator. On the other hand, from the formulas (3.80) which contain derivatives, we get

$$f_0 = \left(\log \frac{\tau_{1,1}}{\tau_{1,0}}\right)' + \frac{t}{3}, \quad f_1 = \left(\log \frac{\tau_{0,0}}{\tau_{1,1}}\right)' + \frac{t}{3}, \quad f_2 = \left(\log \frac{\tau_{1,0}}{\tau_{0,0}}\right)' + \frac{t}{3}. \tag{4.57}$$

Applying $T_1^m T_2^n$ to these, we obtain

$$
\begin{aligned}
T_1^m T_2^n(f_0) &= f_0 + \left(\log \frac{\phi_{m+1,n+1}}{\phi_{m+1,n}}\right)', \\
T_1^m T_2^n(f_1) &= f_1 + \left(\log \frac{\phi_{m,n}}{\phi_{m+1,n+1}}\right)', \\
T_1^m T_2^n(f_2) &= f_2 + \left(\log \frac{\phi_{m+1,n}}{\phi_{m,n}}\right)'.
\end{aligned} \tag{4.58}
$$

The two types of formulas derived above should give the same functions. Hence we see that the following relations hold:

$$
\begin{aligned}
\phi_{m,n}\phi_{m+2,n+1} &= (D_t + f_0)\phi_{m+1,n+1} \cdot \phi_{m+1,n}, \\
\phi_{m+1,n}\phi_{m,n+1} &= (D_t + f_1)\phi_{m,n} \cdot \phi_{m+1,n+1}, \\
\phi_{m+1,n+1}\phi_{m,n-1} &= (D_t + f_2)\phi_{m+1,n} \cdot \phi_{m,n}.
\end{aligned} \tag{4.59}
$$

Whenever we specify a pair of regular triangles that have one edge in common in Figure 4.1, we obtain such relations among the four ϕ-factors corresponding to the vertices of the rhombus. These relations can also be employed as the recurrence relations for determining the ϕ-factors.

For simplicity, let us look only at the direction T_1:

(4.60)

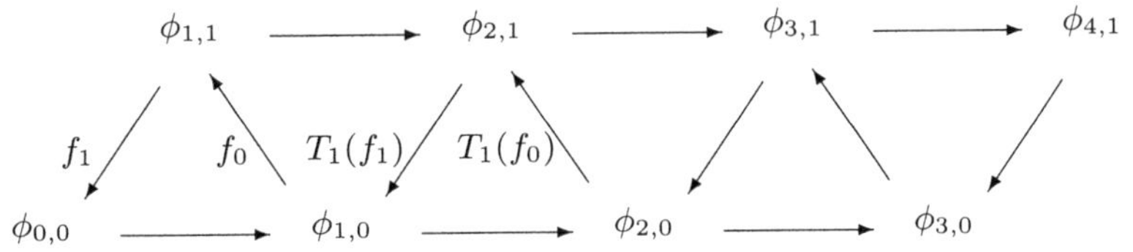

Since the relation $f_0 + f_1 + f_2 = t$ is preserved, we have only to consider the Bäcklund transformations of f_0 and f_1. Then, $T_1^m(f_0)$ and $T_1^m(f_1)$ are expressed as

$$
T_1^m(f_0) = \frac{\phi_{m,0}\,\phi_{m+2,1}}{\phi_{m+1,0}\,\phi_{m+1,1}}, \quad T_1^m(f_1) = \frac{\phi_{m+1,0}\,\phi_{m,1}}{\phi_{m,0}\,\phi_{m+1,1}} \tag{4.61}
$$

by means of the ϕ-factors. Setting

$$
F_m = \phi_{m,0}, \quad G_m = \phi_{m+1,1}, \tag{4.62}
$$

we obtain

$$
T_1^m(f_0) = \frac{F_m G_{m+1}}{F_{m+1} G_m}, \quad T_1^m(f_1) = \frac{F_{m+1} G_{m-1}}{F_m G_m}. \tag{4.63}
$$

The recurrence formulas (4.59) that include differentiation are rewritten into the relations

$$
\begin{aligned}
F_m G_{m+1} &= -(D_t + f_0)F_{m+1} \cdot G_m, \\
F_{m+1} G_{m-1} &= (D_t + f_1)F_m \cdot G_m
\end{aligned} \tag{4.64}
$$

for F_m, G_m. Starting from the initial condition

$$
F_0 = F_1 = 1, \qquad G_0 = 1, \tag{4.65}
$$

we can determine all F_m, G_m $(m = 0, 1, 2, \dots)$ by (4.64):

(4.66)

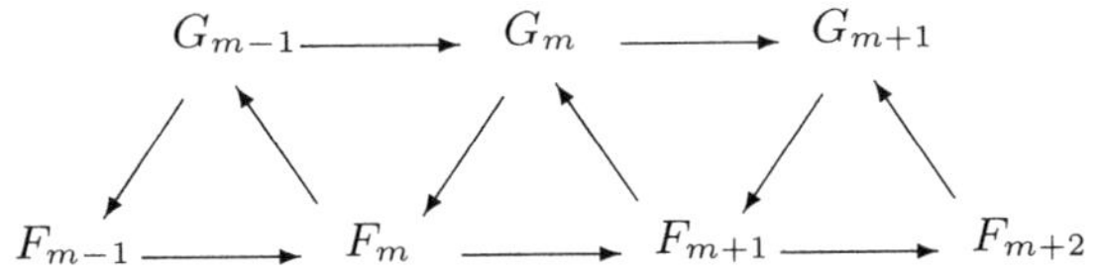

TABLE 4.1. Okamoto Polynomials

$$
\begin{aligned}
&Q_0 = Q_1 = 1, \quad Q_2 = 1 + x^2, \quad Q_3 = 5 + 5x^2 + 5x^4 + x^6,\\
&Q_4 = 175 + 350x^2 + 175x^4 + 140x^6 + 65x^8 + 14x^{10} + x^{12},\\
&Q_5 = 67375 + 134750x^2 + 202125x^4 + 107800x^6 + 42350x^8\\
&\qquad + 20020x^{10} + 8050x^{12} + 2200x^{14} + 355x^{16} + 30x^{18} + x^{20},\\
&\qquad \dots\\
&R_0 = 1, \quad R_1 = x, \quad R_2 = -1 + 2x^2 + x^4,\\
&R_3 = -35x + 14x^5 + 8x^7 + x^9,\\
&R_4 = 1225 - 4900x^2 - 4900x^4 - 980x^6 + 350x^8\\
&\qquad + 420x^{10} + 140x^{12} + 20x^{14} + x^{16},\\
&\qquad \dots
\end{aligned}
$$

We remark that one can also use the Toda equation (4.46) for determining F_m and G_m recursively. In this case we obtain the recurrence relations for F_m and for G_m separately.

Exercise: Derive the Toda equations for F_m, G_m from (4.46).

From the definition we only know that F_m and G_m are rational functions. However, if we carry out explicit computations, we observe that they are actually polynomials in f_j, α_j:

$$
\begin{aligned}
&F_0 = F_1 = 1, \quad F_2 = f_0 f_2 - \alpha_0,\\
&F_3 = f_0^3 f_1 f_2^2 - (\alpha_0 + \alpha_2) f_0^3 f_2 - (\alpha_0 - \alpha_2) f_0^2 f_1 f_2 + \cdots,\\
&G_0 = 1, \quad G_1 = f_0,\\
&G_2 = f_0^2 f_1 f_2 - (\alpha_0 + \alpha_2) f_0^2 - \alpha_0 f_0 f_1 + \alpha_0 f_0 f_2 - \alpha_0^2,\\
&G_3 = f_0^4 f_1^2 f_2^3 + (\alpha_1 - \alpha_2) f_0^4 f_1 f_2^2 + \cdots.
\end{aligned}
\tag{4.67}
$$

If we specialize the f variables to the rational solution

$$
(\alpha_0, \alpha_1, \alpha_2; f_0, f_1, f_2) = \left(\frac{1}{3}, \frac{1}{3}, \frac{1}{3}; \frac{t}{3}, \frac{t}{3}, \frac{t}{3}\right),
\tag{4.68}
$$

then we obtain the following sequences of polynomials in t:

$$
\begin{aligned}
&F_0 = F_1 = 1, \quad F_2 = -\frac{1}{3} + \frac{t^2}{9},\\
&F_3 = -\frac{5}{27} + \frac{5t^2}{81} - \frac{5t^4}{243} + \frac{t^6}{729}, \quad \dots,\\
&G_0 = 1, \quad G_1 = \frac{t}{3}, \quad G_2 = -\frac{1}{9} - \frac{2t^2}{27} + \frac{t^4}{81},\\
&G_3 = -\frac{35t}{243} + \frac{14t^5}{2187} - \frac{8t^7}{6561} + \frac{t^9}{19683}, \quad \dots.
\end{aligned}
\tag{4.69}
$$

Polynomials in these sequences are essentially the *Okamoto polynomials.* To be more precise, by rescaling the variable these F_m and G_m can be transformed into polynomials with integer coefficients. It is the polynomials Q_m normalized as in Table 4.1 that are usually called the Okamoto polynomials.

Consider the Toda equation

$$Q_{m-1}Q_{m+1} = \left(\frac{1}{2}D_x + x^2 + 2m - 1\right) Q_m \cdot Q_m \quad (m \in \mathbb{Z}) \tag{4.70}$$

with the initial condition $Q_0 = Q_1 = 1$. Regarding it as the recurrence relation for $Q_m = Q_m(x)$, we obtain the rational functions Q_m $(m \in \mathbb{Z})$ in x. Although Q_m are rational functions by definition, it is known that they are actually polynomials in x. In fact Q_m is a monic polynomial of degree $m(m-1)$ and the polynomial F_m in (4.69) is a constant multiple of the one obtained from $Q_m(x)$ by the substitution $x = t/\sqrt{-3}$. The polynomials G_m correspond to R_m that are defined by the Toda equation

$$R_{m-1}R_{m+1} = \left(\frac{1}{2}D_x^2 + x^2 + 2m\right) R_m \cdot R_m \quad (m \in \mathbb{Z}) \tag{4.71}$$

with initial condition $R_0 = 1$, $R_1 = x$; $R_m(x)$ is a monic polynomial of degree m^2, and G_m in (4.69) is a constant multiple of $R_m(t/\sqrt{-3})$. The recurrence relations for Q_m, R_m that correspond to (4.64) are given by

$$\begin{aligned} Q_m R_{m+1} &= (D_x + x)\, Q_{m+1} \cdot R_m, \\ Q_{m+1} R_{m-1} &= -\,(D_x + x)\, Q_m \cdot R_m. \end{aligned} \tag{4.72}$$

Remark: It is because of the normalization $\alpha_0+\alpha_1+\alpha_2 = 1$ for the symmetric form that the coefficients of the polynomials F_m and G_m, specialized to the rational solution, contain redundant trivial factors. We could have chosen the normalization so that the Okamoto polynomials Q_m and R_m appear directly in place of F_m and G_m. But we will not say more on this point here.

4.4. Discrete system of type $A_{n-1}^{(1)}$

Recall that the τ-functions on the lattice, as well as the ϕ-factors associated with them, are determined only from the structure of Bäcklund transformations by the affine Weyl group, of type $A_1^{(1)}$ in the case of $P_{\rm II}$, and of type $A_2^{(1)}$ in the case of $P_{\rm IV}$. The Painlevé equation $P_{\rm V}$ admits the action of the affine Weyl group of type $A_3^{(1)}$, although we will not deal with this case in this book. The structure of rational transformations, common to these Painlevé equations, can be generalized naturally to the case of the affine Weyl group of type $A_{n-1}^{(1)}$. In this section we formulate a group of birational transformations of type $A_{n-1}^{(1)}$ for an arbitrary n. Viewed as a system of difference equations, it can be thought of as an extension of discrete Painlevé equations.

Let $\varepsilon_1, \ldots, \varepsilon_n$ be the canonical coordinates of the n-dimensional affine space $V = \mathbb{C}^n$. In the following we fix a constant $\delta \in \mathbb{C}$ and use it as a parameter for measuring distances in the lattice. On these ε variables we define the following transformations $s_1, \ldots, s_{n-1}$:

$$s_i(\varepsilon_i) = \varepsilon_{i+1}, \quad s_i(\varepsilon_{i+1}) = \varepsilon_i, \quad s_i(\varepsilon_j) = \varepsilon_j \quad (j \neq i, i+1). \tag{4.73}$$

Also we define s_0 and π for this case by

$$\begin{aligned} &s_0(\varepsilon_1) = \varepsilon_n + \delta, \quad s_0(\varepsilon_n) = \varepsilon_1 - \delta, \quad s_0(\varepsilon_j) = \varepsilon_j \quad (j \neq 1, n), \\ &\pi(\varepsilon_i) = \varepsilon_{i+1} \quad (i = 1, \ldots, n-1), \quad \pi(\varepsilon_n) = \varepsilon_1 - \delta. \end{aligned} \tag{4.74}$$

If one ignores the presence of δ, then s_i $(i = 1, \dots, n-1)$ and s_0 correspond to the transpositions $(i\ i+1)$ and $(1\ n)$, respectively, and π to the cyclic permutation $(1\,2\,\dots\,n)$. It is useful to extend the indexing set for the ε variables to the whole $\mathbb{Z}$ by the rule

$$\varepsilon_{j+n} = \varepsilon_j - \delta \qquad (j \in \mathbb{Z}). \tag{4.75}$$

With this convention the actions of s_i $(i = 0, 1, \dots, n-1)$ and π take simpler forms:

$$\begin{aligned} &\pi(\varepsilon_j) = \varepsilon_{j+1} && (j \in \mathbb{Z}), \\ &s_i(\varepsilon_j) = \varepsilon_{j+1} && (j \equiv i), \quad s_i(\varepsilon_j) = \varepsilon_{j-1} \quad (j \equiv i+1), \\ &s_i(\varepsilon_j) = \varepsilon_j && (j \not\equiv i, i+1). \end{aligned} \tag{4.76}$$

We now define the *simple roots* $\alpha_0, \alpha_1, \dots, \alpha_{n-1}$ by

$$\alpha_0 = \delta - \varepsilon_1 + \varepsilon_n, \quad \alpha_i = \varepsilon_i - \varepsilon_{i+1} \ (i = 1, \dots n-1), \tag{4.77}$$

so that

$$\alpha_0 + \alpha_1 + \cdots + \alpha_{n-1} = \delta. \tag{4.78}$$

Then the action of s_i $(i = 0, 1, \dots, n-1)$ on these variables is given by

$$\begin{aligned} &s_i(\alpha_i) = -\alpha_i, && s_i(\alpha_j) = \alpha_j + \alpha_i \ (j \equiv i \pm 1), \\ &s_i(\alpha_j) = \alpha_j && (j \not\equiv i, i \pm 1). \end{aligned} \tag{4.79}$$

Here we have used the notation $a \equiv b$ for $a \equiv b \pmod n$, namely $a - b \in n\mathbb{Z}$. We set $\alpha_{i+n} = \alpha_i$ by regarding the indices for α_i as elements of $\mathbb{Z}/n\mathbb{Z}$; then we have

$$\pi(\alpha_i) = \alpha_{i+1} \qquad (i = 0, 1, \dots, n-1). \tag{4.80}$$

(We have $\alpha_i = \varepsilon_i - \varepsilon_{i+1}$ for any i, as we have extended the indexing set for ε variables to $\mathbb{Z}$.) These variables $\alpha_0, \alpha_1, \dots, \alpha_{n-1}$ correspond to the parameters of the symmetric forms of P_{II} and P_{IV}. (In the discussion of P_{II} and P_{IV} we have used the normalization $\delta = 1$.)

The action of s_i on α_j is described by the Cartan matrix of type $A^{(1)}_{n-1}$:

$$s_i(\alpha_j) = \alpha_j - \alpha_i\, a_{ij} \quad (i, j = 0, 1 \dots, n-1). \tag{4.81}$$

The Cartan matrix for this case is given by

$$a_{jj} = 2, \quad a_{ij} = -1 \ (j \equiv i \pm 1), \quad a_{ij} = 0 \ (j \not\equiv i, i \pm 1). \tag{4.82}$$

The corresponding Dynkin diagram is the ring consisting of n nodes $\circ$:

$$A = \begin{bmatrix} 2 & -1 & & & -1 \\ -1 & 2 & -1 & & \\ & -1 & 2 & \ddots & \\ & & \ddots & \ddots & -1 \\ -1 & & & -1 & 2 \end{bmatrix} \tag{4.83}$$

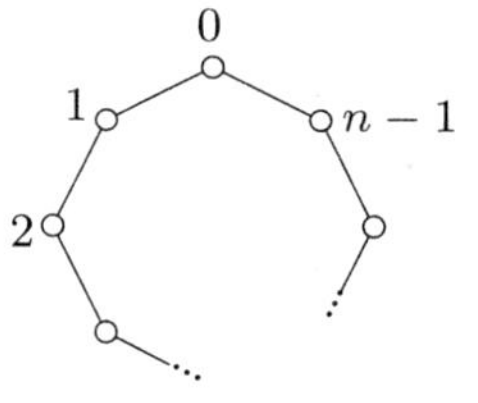

The action of π on α_i corresponds to the rotation of the Dynkin diagram.

The action of $s_0, \dots, s_{n-1}$ on the α variables, as described above, is the canonical realization of the affine Weyl group $W = W(A^{(1)}_{n-1})$ of type $A^{(1)}_{n-1}$. Here, by

$W = W(A_{n-1}^{(1)})$ we mean the group defined by the generators $s_0, s_1, \ldots, s_{n-1}$ and the fundamental relations

$$s_i^2 = 1, \quad (s_i s_j)^2 = 1 \ \ (j \not\equiv i, i \pm 1), \quad (s_i s_j)^3 = 1 \ \ (j \equiv i \pm 1). \tag{4.84}$$

Note that the relation for the product $s_i s_j$ can be alternatively replaced by the braid relation $s_i s_j s_i = s_j s_i s_j$. In what follows, we denote by $\widetilde{W} = \langle\, s_0, s_1, \ldots, s_{n-1}, \pi \,\rangle$ the extension of $W = \langle\, s_0, s_1, \ldots, s_{n-1} \,\rangle$ obtained by adjoining an element π through the relations

$$\pi\, s_i = s_{i+1}\, \pi \qquad (i = 0, 1, \ldots, n-1). \tag{4.85}$$

Here we have regarded the indices for s_i as elements of $\mathbb{Z}/n\mathbb{Z}$ as before. It is easily seen that these relations hold in the realization as mentioned above.

> **Remark:** When one extends $W = W(A_{n-1}^{(1)})$ by the diagram rotation, it is probably standard to introduce π with it being subjected to the relation $\pi^n = 1$. It is in order to simplify the description in later arguments that we have not imposed this relation in the definition of $\widetilde{W}$. This corresponds to considering the Lie algebra $\mathfrak{gl}_n$ instead of $\mathfrak{sl}_n$.

In the group $\widetilde{W} = \langle\, s_0, s_1, \ldots, s_{n-1}, \pi \,\rangle$ defined by the fundamental relations (4.84) and (4.85), we define the n elements $T_1, \ldots, T_n$ by

$$\begin{aligned} T_1 &= \pi s_{n-1} s_{n-2} \cdots s_1, \\ T_2 &= s_1 \pi s_{n-1} s_{n-2} \cdots s_2, \\ &\cdots \\ T_n &= s_{n-1} s_{n-2} \cdots s_1 \pi. \end{aligned} \tag{4.86}$$

Then we have

LEMMA 4.7. *Under the fundamental relations* (4.84) *and* (4.85),
(1) $T_1, \ldots, T_n$ *commute with each other* :

$$T_i T_j = T_j T_i \qquad (i, j = 1, \ldots, n). \tag{4.87}$$

(2) *The commutation relations between* s_i $(i = 0, \ldots, n-1)$, π *and* T_j *are given by*

$$s_i\, T_j = T_{\sigma_i(j)}\, s_i, \quad \pi\, T_j = T_{\rho(j)} \pi \qquad (j = 1, \ldots, n). \tag{4.88}$$

In this lemma we have used the following notations of transpositions and cyclic permutations:

$$\sigma_0 = (1\, n), \quad \sigma_i = (i\ i+1) \ \ (i = 1, \ldots, n-1); \quad \rho = (1 2 \ldots n). \tag{4.89}$$

The elements T_i $(i = 1, \ldots, n)$ defined above act on the ε variables as

$$T_i(\varepsilon_i) = \varepsilon_i - \delta, \quad T_i(\varepsilon_j) = \varepsilon_j \quad (j \neq i). \tag{4.90}$$

Namely, in the affine space $V = \mathbb{C}^n$ having the ε variables as coordinates each T_i corresponds to the parallel translation by δ in the direction of the i-th coordinate. It is obvious that the relations appearing in the lemma are valid in this realization. The lemma above is then a consequence of the fact that the realization of $\widetilde{W}$ on the ε variables is faithful (that is, all the relations are reduced to (4.84) and (4.85)). Here we will not get into the proof of the faithfulness. It is also possible to prove the lemma itself directly from the relations (4.84), (4.85) and the definition (4.86).

It would be a challenge for the reader to try such a proof. By the definition (4.86), one can also verify that

$$T_1 T_2 \cdots T_r = (\pi s_{n-1} \cdots s_r)^r \qquad (r = 1, \ldots, n). \tag{4.91}$$

Since in particular $T_1 \cdots T_n = \pi^n$, the product $T_1 \cdots T_n$ gives the identity transformation in a realization such that π^n acts as the identity transformation.

Lemma 4.7 means that, given transformations $s_0, s_1, \ldots, s_{n-1}, \pi$ satisfying the relations (4.84), (4.85), by defining T_i as in (4.86) one can construct a family of transformations which are mutually commutative, and equivariant with respect to the actions of s_i, π. Namely, if one can realize the extended affine Weyl group $\widetilde{W}$ as a group of transformations, then one automatically obtains a discrete integrable system (system of difference equations). In what follows we introduce the f variables, corresponding to those in the symmetric forms of P_{II}, P_{IV}, and propose a way to realize $\widetilde{W}$ as a group of rational transformations on the space having the ε variables and the f variables as coordinates. This realization can be extended further to the level of τ functions and used to construct a discrete system on the lattice of type $A^{(1)}_{n-1}$. In this section we deal with a *simple* version which gives an extension of the case of P_{IV}; here we assume that $n \geq 3$, excluding the type $A^{(1)}_1$. A more general version will be discussed from the next chapter onward.

Taking the n variables $f_0, \ldots, f_{n-1}$, we define the transformations $s_0, \ldots, s_{n-1}$ and π of these variables as follows: For any $i, j = 0, 1, \ldots, n-1$,

$$\begin{aligned} &s_i(f_j) = f_j \pm \frac{\alpha_i}{f_i} \quad (j \equiv i \pm 1), \quad s_i(f_j) = f_j \quad (j \not\equiv i \pm 1), \\ &\pi(f_j) = f_{j+1}. \end{aligned} \tag{4.92}$$

As before, we regard the indices as elements of $\mathbb{Z}/n\mathbb{Z}$. We define the matrix $U = (u_{ij})_{i,j=0}^{n-1}$ by

$$u_{ij} = \pm 1 \quad (j \equiv i \pm 1), \quad u_{ij} = 0 \quad (j \not\equiv i \pm 1), \tag{4.93}$$

and make use of the Poisson bracket $\{f_i, f_j\} = u_{ij}$. Then the actions described above are expressed in the form

$$s_i(f_j) = f_j + \frac{\alpha_i}{f_i} u_{ij} = f_j + \frac{\alpha_i}{f_i}\{f_i, f_j\} \quad (i, j = 0, \ldots, n-1) \tag{4.94}$$

in terms of the Poisson bracket. The matrix U represents a orientation of the Dynkin diagram:

$$U = \begin{bmatrix} 0 & 1 & & & -1 \\ -1 & 0 & 1 & & \\ & -1 & 0 & \ddots & \\ & & \ddots & \ddots & 1 \\ 1 & & & -1 & 0 \end{bmatrix} \tag{4.95}$$

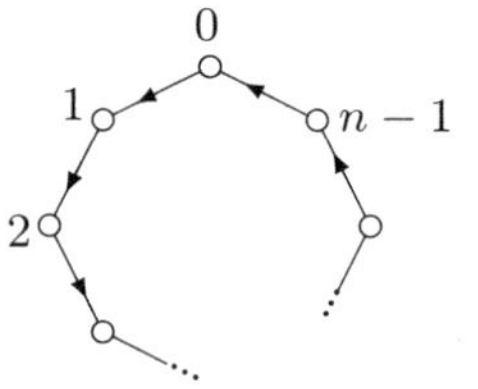

Similarly, we introduce the set of n τ-*functions* $\tau_0, \ldots, \tau_{n-1}$ (τ variables) and define

$$s_i(\tau_j) = \tau_j \quad (i \neq j), \quad s_j(\tau_j) = f_j \, \frac{\tau_{j-1}\tau_{j+1}}{\tau_j}, \quad \pi(\tau_j) = \tau_{j+1} \tag{4.96}$$

(with indices in $\mathbb{Z}/n\mathbb{Z}$). In this definition we have already built in the multiplicative formula

$$f_j = \frac{\tau_j\ s_j(\tau_j)}{\tau_{j-1}\tau_{j+1}} = \frac{\tau_j\ s_j(\tau_j)}{\prod_{i\neq j}\tau_i^{-a_{ij}}} \qquad (j = 0, 1, \dots, n-1). \tag{4.97}$$

THEOREM 4.8. *For $n \geq 3$, define the transformations $s_0, s_1, \dots, s_{n-1}$, π for the variables ε_j $(j = 1, \dots, n)$, f_j, τ_j $(j = 0, 1, \dots, n-1)$ by* (4.73), (4.74) *and* (4.92), (4.96) *respectively. Then $s_0, s_1, \dots, s_{n-1}$ and π satisfy the relations* (4.84), (4.85). *Namely, they give a realization of the extension $\widetilde{W}$ of the affine Weyl group of type $A^{(1)}_{n-1}$ as a group of birational transformations.*

We remark that these transformations are closed within the variables α_j, f_j, τ_j; as transformations of these variables they satisfy $\pi^n = 1$ and hence $T_1 \cdots T_n = 1$. We omit the proof of this theorem, since it is essentially the same as in the case of P_{IV}.

We now try to write down the explicit difference system arising from this realization. For example, T_1 induces the transformation

$$T_1(\alpha_0) = \alpha_0 + \delta, \quad T_1(\alpha_1) = \alpha_1 - \delta, \quad T_1(\alpha_j) = \alpha_j \quad (j \neq 0, 1) \tag{4.98}$$

of the α variables. By applying the product $T_1 = \pi s_{n-1} \cdots s_1$ one by one, we have

$$\begin{aligned} f_0 &\overset{s_1}{\longrightarrow} f_0 - \frac{\alpha_1}{f_1} \overset{s_{n-2}\cdots s_2}{\longrightarrow} f_0 - s_{n-2}\cdots s_2(\frac{\alpha_1}{f_1}) \\ &\overset{s_{n-1}}{\longrightarrow} f_0 + \frac{\alpha_{n-1}}{f_{n-1}} - s_{n-1}\cdots s_2(\frac{\alpha_1}{f_1}). \end{aligned} \tag{4.99}$$

Finally, rotating the indices by π we obtain

$$T_1(f_0) = f_1 + \frac{\alpha_0}{f_0} - s_0 s_{n-1}\cdots s_3(\frac{\alpha_2}{f_2}). \tag{4.100}$$

In view of this formula let us define

$$g_{i,j} = s_{j-1}\cdots s_{i+1}(\frac{\alpha_i}{f_i}) \tag{4.101}$$

for $i, j \in \mathbb{Z}$ with $i < j$; as we apply s_k, we increase the index k one by one in $\mathbb{Z}/n\mathbb{Z}$. The repetition of transformations as in (4.92) then produces continued fractions in the following form:

$$\begin{aligned} g_{i,j} &= \cfrac{\beta_{i,j}}{f_i - \cfrac{\beta_{i+1,j}}{f_{i+1} - \cfrac{\beta_{i+2,j}}{\ddots \, \overline{\;f_{j-2} - \cfrac{\beta_{j-1,j}}{f_{j-1}}\;}}}} \\ &= \frac{\beta_{i,j}|}{|\ f_i} - \frac{\beta_{i+1,j}|}{|\ f_{i+1}} - \cdots - \frac{\beta_{j-1,j}|}{|\ f_{j-1}}. \end{aligned} \tag{4.102}$$

Here we have set $\beta_{i,j} = s_{j-1}\cdots s_{i+1}(\alpha_i)$. Note that for $i < j < n + i$

$$\beta_{i,j} = \alpha_i + \alpha_{i+1} + \cdots + \alpha_{j-1} = \varepsilon_i - \varepsilon_j. \tag{4.103}$$

With g_{ij} defined as above, the action of T_1 on f_j is expressed as follows:

$$
\begin{aligned}
T_1(f_0) &= f_1 + g_{0,1} - g_{2,n+1},\\
T_1(f_1) &= f_2 - g_{3,n+1},\\
T_1(f_2) &= f_3 + g_{2,n+1} - g_{4,n+1},\\
&\cdots\\
T_1(f_{n-2}) &= f_{n-1} + g_{n-2,n+1} - g_{n,n+1},\\
T_1(f_{n-1}) &= f_0 + g_{n-1,n+1}.
\end{aligned}
\tag{4.104}
$$

Since $\pi T_i = T_{i+1}\pi$, the action of each $T_2, \ldots, T_n$ is obtained by rotating the indices. Recall that T_k plays the role of shifting the variable ε_k by $-\delta$. It means that the numerators built in the continued fractions $g_{i,j}$ contain the independent variables.

When we need to write down the difference system explicitly, we apply the following procedure. For each $i = 1, \ldots, n$ and $j = 0, \ldots, n-1$, there exists a rational function F_{ij} in α_k, f_k such that

$$T_i(f_j) = F_{ij}(\alpha_0, \ldots, \alpha_{n-1}; f_0, \ldots, f_{n-1}). \tag{4.105}$$

(In the argument above, we gave a representation of F_{ij} in terms of continued fractions.) For each multi-index $\nu = (\nu_1, \ldots, \nu_n) \in \mathbb{Z}^n$ we set

$$f_j(\nu) = T^{\nu}(f_j), \quad T^{\nu} = T_1^{\nu_1} \cdots T_n^{\nu_n}. \tag{4.106}$$

Then applying T^{ν} to (4.105) we obtain

$$f_j(\nu + \boldsymbol{e}_i) = F_{ij}(\alpha_0(\nu), \ldots, \alpha_{n-1}(\nu); f_0(\nu), \ldots, f_{n-1}(\nu)), \tag{4.107}$$

where $\boldsymbol{e}_i$ is the i-th unit vector of $\mathbb{Z}^n$. Note also that the shifted α variables $\alpha_j(\nu) = T^{\nu}(\alpha_j)$ are given by

$$
\begin{aligned}
&\alpha_0(\nu) = \alpha_0 - (\nu_n - \nu_1)\delta, \quad \alpha_1(\nu) = \alpha_1 - (\nu_1 - \nu_2)\delta,\\
&\ldots, \quad \alpha_{n-1}(\nu) = \alpha_{n-1} - (\nu_{n-1} - \nu_n)\delta.
\end{aligned}
\tag{4.108}
$$

Formula (4.107) provides us with a representation of the discrete system of type $A^{(1)}_{n-1}$ as a system of difference equations.

As we have seen in (4.36), (4.37) in the case of $A^{(1)}_2$, continued fractions F_{ij}, constructed by means of the action of affine Weyl groups, have a nice factorization. As we will see in the next section, the factorization of F_{ij}, as well as that of general $T^{\nu}(f_j)$, shows a characteristic pattern that can be described in terms of the ϕ-factors. By using the determinant formula of Jacobi-Trudi type, which will be discussed in the next chapter, it is also possible to determine the ϕ-factors themselves explicitly as polynomials in α_k and f_k.

4.5. τ-Functions on the A_{n-1} lattice

As we have done with $P_{\rm II}$ and $P_{\rm IV}$, we now formulate the τ-functions on the lattice and ϕ-factors in the framework of the discrete system of type $A^{(1)}_{n-1}$.

We first remark that for any element w of $\widetilde{W} = \langle s_0, \ldots, s_{n-1}, \pi \rangle$ and for any $m = 0, \ldots, n-1$, there exists a unique rational function $\phi_{w,m}$ in the variables α_j, f_j such that

$$w(\tau_m) = \phi_{w,m}\, \tau_0^{k_0} \cdots \tau_{n-1}^{k_{n-1}}, \tag{4.109}$$

where $k_0, \dots, k_{n-1} \in \mathbb{Z}$ are integers to be determined from w and m. When $w' = s_i w$, from the formula above we have

$$w'(\tau_m) = s_i(\phi_{w,m})\, f_i^{k_i}\, \tau_0^{k'_0} \cdots \tau_{n-1}^{k'_{n-1}}, \tag{4.110}$$

where s_i has picked up the exponent k_i for τ_i. When $w' = \pi w$, π simply rotates the exponents $k_0, \dots, k_{n-1}$. So, the rational functions $\phi_{w,m}$ should satisfy the relations

$$\phi_{s_i w,m} = s_i(\phi_{w,m})\, f_i^{k_i} \quad (i = 0, \dots, n-1), \qquad \phi_{\pi w,m} = \pi(\phi_{w,m}). \tag{4.111}$$

This computation also shows that the general $\phi_{w,m}$ are determined starting from $\phi_{1,m} = 1$. We call $\phi_{w,m}$ the *ϕ-factor* for the τ-function $w(\tau_m)$.

We now consider the ϕ-factors in the case where w is the product of the shift operators $T_1, \dots, T_n$. Setting $L = \mathbb{Z}^n$, for each multi-index $\nu = (\nu_1, \dots, \nu_n) \in L$, we define

$$\tau_\nu = T^\nu(\tau_0), \quad T^\nu = T_1^{\nu_1} \cdots T_n^{\nu_n}, \tag{4.112}$$

and denote by ϕ_ν the ϕ-factor for τ_ν. As to the actions of $T_1, \dots, T_n$ on τ_0, one can verify the following by the definition.

LEMMA 4.9. (1) $T_1 \cdots T_i(\tau_0) = \tau_i \quad (i = 1, \dots, n)$. *In particular* $T_1 \cdots T_n(\tau_0) = \tau_0$.
(2) $T_1 T_n^{-1}(\tau_0) = s_0(\tau_0) = f_0 \dfrac{\tau_1 \tau_{n-1}}{\tau_0}$.

Since $T_1 \cdots T_n(\tau_0) = \tau_0$, τ_ν are defined actually on the lattice $P = L/\mathbb{Z}(\boldsymbol{e}_1 + \cdots + \boldsymbol{e}_n)$ of rank $n-1$, where $\boldsymbol{e}_1, \dots, \boldsymbol{e}_n$ stand for the canonical basis of $L = \mathbb{Z}^n$. For each $i = 0, \dots, n-1$, set

$$\varpi_i = \boldsymbol{e}_1 + \cdots + \boldsymbol{e}_i \qquad \mathrm{mod} \quad \mathbb{Z}(\boldsymbol{e}_1 + \cdots + \boldsymbol{e}_n). \tag{4.113}$$

Then the lemma above implies

$$\tau_i = \tau_{\varpi_i} \quad (i = 0, \dots, n-1), \quad s_0(\tau_0) = \tau_{\boldsymbol{e}_1 - \boldsymbol{e}_n} = \tau_{\varpi_1 + \varpi_{n-1}}. \tag{4.114}$$

In the language of root systems, P is called the *weight lattice* of type $A_{n-1}^{(1)}$ and $\varpi_1, \dots, \varpi_{n-1}$ the *fundamental weights.* (It is convenient to set $\varpi_0 = 0$, considering the indices for ϖ_i in $\mathbb{Z}/n\mathbb{Z}$.)

On the lattice L (or P), the symmetric group $\mathfrak{S}_n$ acts through the permutation of indices for the canonical basis (permutation representation):

$$\sigma.\boldsymbol{e}_j = \boldsymbol{e}_{\sigma(j)} \qquad (j = 1, \dots, n). \tag{4.115}$$

With the notation of (4.89), from Lemma 4.7 we have

$$s_i T^\nu = T^{\sigma_i.\nu} s_i \quad (i = 0, \dots, n-1), \quad \pi T^\nu = T^{\rho.\nu}\pi \tag{4.116}$$

for each $\nu \in L$. Define the action of $s_0, \dots, s_{n-1}, \pi$ on L by affine transformations as

$$\begin{aligned} s_0(\nu) &= \sigma_0.\nu + \boldsymbol{e}_1 - \boldsymbol{e}_n = (\nu_n + 1, \nu_2, \dots, \nu_{n-1}, \nu_1 - 1), \\ s_i(\nu) &= \sigma_i.\nu = (\nu_1, \dots, \nu_{i+1}, \nu_i, \dots, \nu_n) \qquad (i = 1, \dots, n-1), \\ \pi(\nu) &= \rho.\nu + \boldsymbol{e}_1 = (\nu_n + 1, \nu_1, \dots, \nu_{n-1}), \end{aligned} \tag{4.117}$$

for each $\nu = (\nu_1, \dots, \nu_n) \in L$. Then we obtain the action $\widetilde{W} = \langle s_0, \dots, s_{n-1}, \pi \rangle$ on L, through which T^ν $(\nu \in L)$ give rise to the parallel translations:

$$T^\nu(\mu) = \mu + \nu \qquad (\mu \in L). \tag{4.118}$$

(This action is dual to the representation of $\widetilde{W}$ with $\delta = 1$ on the lattice generated by ε_j.) By combining Lemma 4.9 with (4.116) we obtain the following proposition.

PROPOSITION 4.10. *The action of the extended affine Weyl group $\widetilde{W}$ on the τ-functions is compatible with its action on the lattice L:*

$$w(\tau_\nu) = \tau_{w(\nu)} \qquad (w \in \widetilde{W}, \quad \nu \in L). \tag{4.119}$$

In order to study the ϕ-factors for the transforms of τ-functions, it is convenient to introduce the variables

$$z_1 = \frac{\tau_1}{\tau_0}, \quad \ldots, \quad z_{n-1} = \frac{\tau_{n-1}}{\tau_{n-2}}, \quad z_n = \frac{\tau_0}{\tau_{n-1}} \tag{4.120}$$

for $i = 1, \ldots, n$, so that we can separate $\tau_0, \tau_1, \ldots, \tau_{n-1}$ into τ_0 and $z_1, \ldots, z_n$. Then the actions of s_i, π on z_j are given by

$$\begin{gathered} s_i(z_i) = f_i z_{i+1}, \quad s_i(z_{i+1}) = \frac{1}{f_i} z_i, \\ s_i(z_j) = z_j \quad (j \neq i, i+1), \quad \pi(z_j) = z_{j+1}. \end{gathered} \tag{4.121}$$

(Again we consider the indices for z_j as elements of $\mathbb{Z}/n\mathbb{Z}$.) From the action of $\mathfrak{S}_n$ on L we can read off how z_j are transformed. With the notation $z^\nu = z_1^{\nu_1} \cdots z_n^{\nu_n}$ of multi-indices for $\nu = (\nu_1, \ldots, \nu_n) \in L$, we have in general

$$s_i(z^\nu) = f_i^{\nu_i - \nu_{i+1}} z^{\sigma_i . \nu}, \quad \pi(z^\nu) = z^{\rho . \nu}. \tag{4.122}$$

Combining this fact with the fact that $\sigma_{i-1} \cdots \sigma_1 \rho \, \sigma_{n-1} \cdots \sigma_i = 1$, we see that $T_i z^\nu$ is expressed as $T_i(z^\nu) = \psi_{i,\nu} z^\nu$ with an appropriate rational function $\psi_{i,\nu} \neq 0$ in the α variables and the f variables. Furthermore we see that there exists a set of rational functions $\psi_{\mu,\nu}$ in the α variables and the f variables such that

$$T^\mu(z^\nu) = \psi_{\mu,\nu} z^\nu \quad (\mu, \nu \in L) \tag{4.123}$$

in general. Since Lemma 4.9 implies

$$T_1 \cdots T_i(\tau_0) = \tau_i = \tau_0 \, z_1 \cdots z_i \qquad (i = 1, \ldots, n), \tag{4.124}$$

from (4.123) we finally obtain the following proposition.

PROPOSITION 4.11. *For each $\nu \in L$,*

$$\tau_\nu = \phi_\nu \, \tau_0 \, z^\nu = \phi_\nu \, \tau_0^{1-\nu_1+\nu_n} \tau_1^{\nu_1 - \nu_2} \cdots \tau_{n-1}^{\nu_{n-1} - \nu_n}. \tag{4.125}$$

It is known that the ϕ-factors ϕ_ν are polynomials in α_j, f_j with integer coefficients and have explicit formulas in terms of determinants of Jacobi-Trudi type. Such properties of the ϕ-factors will be discussed in detail from the next chapter onward.

The formula for representing the f variables in terms of τ-functions implies

$$f_j = \frac{\tau_j \, s_j(\tau_j)}{\tau_{j-1}\tau_{j+1}} = \frac{\tau_{\varpi_j} \, \tau_{\varpi_{j-1} - \varpi_j + \varpi_{j+1}}}{\tau_{\varpi_{j-1}} \tau_{\varpi_{j+1}}}. \tag{4.126}$$

In fact from $\tau_j = \tau_{\varpi_j}$ and

$$s_i(\varpi_j) = \varpi_j \quad (i \neq j), \quad s_j(\varpi_j) = \varpi_{j-1} - \varpi_j + \varpi_{j+1}, \tag{4.127}$$

we have

$$s_j(\tau_j) = \tau_{s_i(\varpi_j)} = \tau_{\varpi_{j-1}-\varpi_j+\varpi_{j+1}}. \tag{4.128}$$

From this we obtain the multiplicative formula for the transforms of the f variables:

$$\begin{aligned} T^\nu(f_j) &= \frac{\tau_{\nu+\varpi_j}\,\tau_{\nu+\varpi_{j-1}-\varpi_j+\varpi_{j+1}}}{\tau_{\nu+\varpi_{j-1}}\,\tau_{\nu+\varpi_{j+1}}} \\ &= \frac{\phi_{\nu+\varpi_j}\,\phi_{\nu+\varpi_{j-1}-\varpi_j+\varpi_{j+1}}}{\phi_{\nu+\varpi_{j-1}}\,\phi_{\nu+\varpi_{j+1}}}. \end{aligned} \tag{4.129}$$

Here the factors containing τ_0 and $z_1, \ldots, z_n$ have disappeared in the same way as we have seen before. This formula for $T^\nu(f_j)$ can be thought of as that for representing the general solution of the discrete system, which we have written down in the previous section, in terms of its initial values. If one computes $T^\nu(f_j)$ by decomposing T^ν as a combination of $s_0, s_1, \ldots, s_{n-1}, \pi$, then it should give rise to a complicated continued fraction. The formula presented above, however, implies that it is expressed multiplicatively by means of four special polynomials. We remark that the multiplicative formula above can be equivalently written as

$$T^{\nu-\varpi_{j-1}}(f_j) = \frac{\tau_{\nu+\mathbf{e}_j}\,\tau_{\nu+\mathbf{e}_{j+1}}}{\tau_\nu\,\tau_{\nu+\mathbf{e}_j+\mathbf{e}_{j+1}}} = \frac{\phi_{\nu+\mathbf{e}_j}\,\phi_{\nu+\mathbf{e}_{j+1}}}{\phi_\nu\,\phi_{\nu+\mathbf{e}_j+\mathbf{e}_{j+1}}}. \tag{4.130}$$

In a similar way as we have done with P_{IV}, we can also translate the action of s_i on f_j into the language of τ-functions. The result is the bilinear relation

$$\tau_{\nu+\mathbf{e}_i}\tau_{\nu+\mathbf{e}_j+\mathbf{e}_k} - \tau_{\nu+\mathbf{e}_j}\tau_{\nu+\mathbf{e}_i+\mathbf{e}_k} = (\varepsilon_i - \varepsilon_j - \nu_i + \nu_j)\tau_\nu\tau_{\nu+\mathbf{e}_i+\mathbf{e}_j+\mathbf{e}_k} \tag{4.131}$$

for three distinct indices $i, j, k = 1, \ldots, n$. From this we obtain the *Hirota-Miwa equation*

$$\begin{aligned} &(\varepsilon_j - \varepsilon_k - \nu_j + \nu_k)\tau_{\nu+\mathbf{e}_i}\tau_{\nu+\mathbf{e}_j+\mathbf{e}_k} \\ &-(\varepsilon_i - \varepsilon_k - \nu_i + \nu_k)\tau_{\nu+\mathbf{e}_j}\tau_{\nu+\mathbf{e}_i+\mathbf{e}_k} \\ &+(\varepsilon_i - \varepsilon_j - \nu_i + \nu_j)\tau_{\nu+\mathbf{e}_k}\tau_{\nu+\mathbf{e}_i+\mathbf{e}_j} = 0. \end{aligned} \tag{4.132}$$

By rewriting this by means of ϕ-factors we see that the polynomials ϕ_ν again satisfy the same bilinear relations. The Jacobi-Trudi formula, which will be discussed in the next chapter, implies that the Hirota-Miwa equation above has a particular solution expressed in terms of determinants.

As for the discrete system of type $A_{n-1}^{(1)}$ $(n \geq 3)$, discussed in these two sections, it is known that there exists a system of ordinary differential equations for which the discrete system plays the role of Bäcklund transformations; such a differential system could be called the (higher order) Painlevé equations of type $A_{n-1}^{(1)}$. The following is an example of such a differential equation of type $A_3^{(1)}$ $(n = 4)$:

$$\begin{aligned} f_0' &= f_0(f_1f_2 - f_2f_3) + (\tfrac{\delta}{2} - \alpha_2)f_0 + \alpha_0 f_2, \\ f_1' &= f_1(f_2f_3 - f_3f_0) + (\tfrac{\delta}{2} - \alpha_3)f_1 + \alpha_1 f_3, \\ f_2' &= f_2(f_3f_0 - f_0f_1) + (\tfrac{\delta}{2} - \alpha_0)f_2 + \alpha_2 f_0, \\ f_3' &= f_3(f_0f_1 - f_1f_2) + (\tfrac{\delta}{2} - \alpha_1)f_3 + \alpha_3 f_1. \end{aligned} \tag{4.133}$$

(Observe the rotational symmetry with respect to the indices.) Although it is a system of fourth order, one can find elementary integrals and normalize the variables so that

$$f_0 + f_2 = f_1 + f_3 = t^{\frac{1}{2}}, \quad ' = t\frac{d}{dt} \tag{4.134}$$

(we have set $\delta = 1$). In this sense the differential system above is essentially of second order. Setting $f_0 = t^{\frac{1}{2}}(1-y)^{-1}$, we can transform the system into a single differential equation of second order for y, which in fact coincides with the fifth Painlevé equation $P_{\rm V}$ listed in Table 0.1 of the preface. (The parameters $\alpha, \beta, \gamma, \delta$ in the table are given by $\alpha_0^2/2, \alpha_1^2/2, -\alpha_1 + \alpha_3, -1/2$ respectively.) In this sense the discrete system of type $A_3^{(1)}$ describes the symmetry of $P_{\rm V}$ under the Bäcklund transformations. In the case of $A_4^{(1)}$ ($n = 5$) we have a differential system somewhat similar to the symmetric form of $P_{\rm IV}$ of type $A_2^{(1)}$:

$$\begin{aligned} f_0' &= f_0(f_1 - f_2 + f_3 - f_4) + \alpha_0, \\ f_1' &= f_1(f_2 - f_3 + f_4 - f_0) + \alpha_1, \\ f_2' &= f_2(f_3 - f_4 + f_0 - f_1) + \alpha_2, \\ f_3' &= f_3(f_4 - f_0 + f_1 - f_2) + \alpha_3, \\ f_4' &= f_4(f_0 - f_1 + f_2 - f_3) + \alpha_4. \end{aligned} \tag{4.135}$$

This is again a variation of the so-called *hungry* Lotka-Volterra equation. Under the normalization

$$f_0 + f_1 + f_2 + f_3 + f_4 = t, \quad ' = \frac{d}{dt}, \tag{4.136}$$

we see that the differential system above is essentially of fourth order. We remark that by setting $\alpha_3 = \alpha_4 = 0$, $f_3 = f_4 = 0$, it specializes to the symmetric form of $P_{\rm IV}$. Such systems of ordinary differential equations are known for general $A_{n-1}^{(1)}$ although they have slightly different structures according to the parity of n.

CHAPTER 5

Jacobi-Trudi Formula

In the previous chapter, we gave a formulation of ϕ-factors associated with the transforms of τ-functions, and made an observation that ϕ-factors are polynomials in α_j, f_j. As a matter of fact, ϕ-factors are expressed as certain determinants with polynomial entries. This formula, discovered by Yasuhiko Yamada, is called below the formula of Jacobi-Trudi type, because of its characteristic structure of determinants. In this chapter we explain what the Jacobi-Trudi formula is for the ϕ-factors.

5.1. Young diagrams and Maya diagrams

In this section we introduce the terminology of *Young diagrams* and *Maya diagrams*; these are preliminaries for the formulas of Jacobi-Trudi type.

A non-increasing sequence of integers

$$\lambda = (\lambda_1, \lambda_2, \ldots); \quad \lambda_1 \geq \lambda_2 \geq \cdots \tag{5.1}$$

is called a *partition* if $\lambda_i = 0$ for all sufficiently large i; we set $|\lambda| = \lambda_1 + \lambda_2 + \cdots$. The phrase "partition" comes from the fact that such a λ corresponds precisely to a way of expressing the natural number $|\lambda|$ as the sum of natural numbers. For instance, there are five partitions λ such that $|\lambda| = 4$:

$$\lambda: \quad (4), \quad (3,1), \quad (2,2), \quad (2,1,1), \quad (1,1,1,1). \tag{5.2}$$

In writing down partitions explicitly, we usually omit the tail consisting only of 0's. The five listed above correspond to the five ways of expressing 4 as the sum of natural numbers:

$$4 = 3+1 = 2+2 = 2+1+1 = 1+1+1+1. \tag{5.3}$$

We denote by $\ell(\lambda)$ the number of nonzero entries of $\lambda = (\lambda_1, \lambda_2, \ldots)$; in the examples above, $\ell(\lambda)$ is $1, 2, 2, 3, 4$, respectively. In order to visualize partitions, we use the so-called *Young diagrams*:

(5.4) Y :

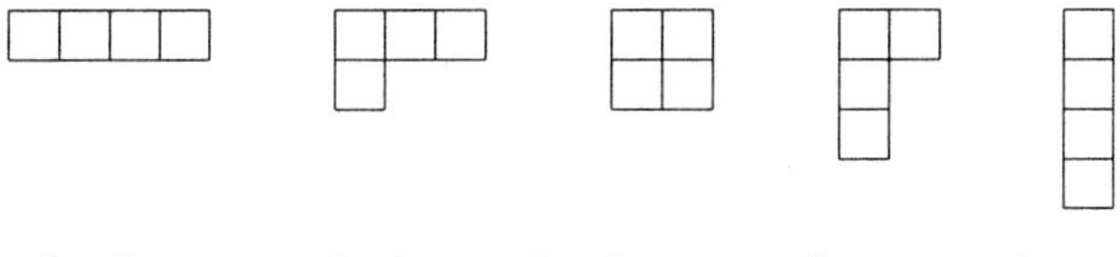

We put λ_1 boxes in the first row, λ_2 boxes in the second row, and so on, in such a way that the boxes are all packed to the left in each row. The number $|Y|$ of boxes in the Young diagram Y is the same as $|\lambda| = \lambda_1 + \lambda_2 + \cdots$, and $\ell(\lambda)$ represents the vertical height of Y.

Exercise: We denote by $p(n)$ the number of partitions λ such that $|\lambda| = n$; $p(n)$ is called the n-th *partition number*. Prove that a generating function for the partition numbers is given by

$$\sum_{n=0}^{\infty} p(n) q^n = \frac{1}{\prod_{k=1}^{\infty}(1-q^k)}. \tag{5.5}$$

It is sometimes more convenient to use an alternative representation of the Young diagram, called the *Maya diagram*. Consider the infinite sequence of boxes labeled by the set of integers $\mathbb{Z}$.

(5.6)

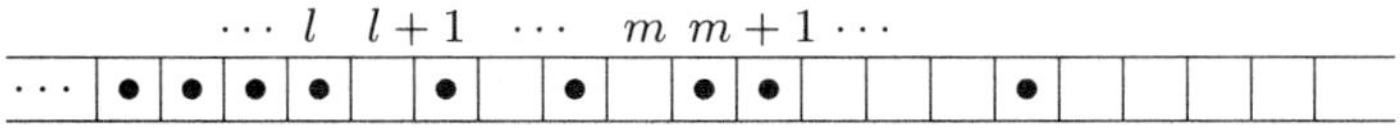

A *Maya diagram* is a way of putting balls • into some of the boxes. We impose the following three restrictions on the arrangement of balls: (1) each box, if not empty, is allowed to contain only one ball; (2) boxes located sufficiently left are all occupied by a ball; (3) boxes located sufficiently right are all empty. (Instead of boxes and balls, you may wish to use white boxes and black boxes.) Given a Maya diagram, one can move some of the balls so that all the boxes are packed to the left; then one can specify a unique $m \in \mathbb{Z}$ such that the boxes labeled by a number $\leq m$ are all occupied, and that those labeled by a number $> m$ are all empty. With this m determined as above, the original Maya diagram is said to have *index* m, and the Maya diagram with all the balls packed to the left is called a *ground state* of index m:

(5.7)

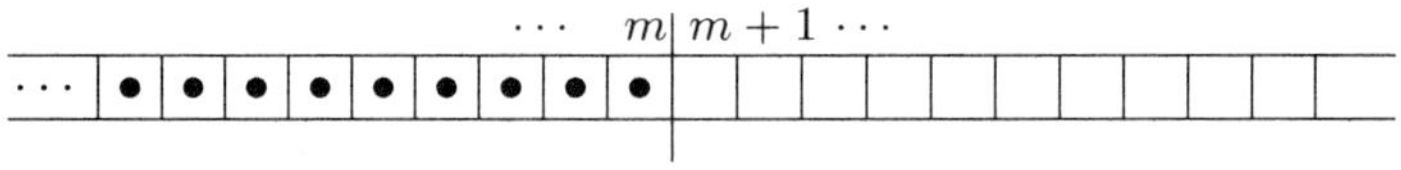

Let us denote by K the set of labels of all nonempty boxes. Then a Maya diagram is identified with a subset K of $\mathbb{Z}$ such that

$$K \cap \mathbb{Z}_{\leq l} = \mathbb{Z}_{\leq l} \quad (l \ll 0), \qquad K \cap \mathbb{Z}_{>n} = \emptyset \quad (n \gg 0). \tag{5.8}$$

Hereafter, we use the expression "$\cdots$ for $l \ll 0$" in place of "There exists an $l_0 < 0$ such that $\ldots$ for all l with $l \leq l_0$". Note also that the subset $\mathbb{Z}_{\leq m}$ of $\mathbb{Z}$ represents the ground state of index m.

Let K be a Maya diagram of index m. Then one can find a sequence of integers k_i $(i \in \mathbb{Z}_{\leq m})$ indexed by $\mathbb{Z}_{\leq m}$ such that

$$K = \{k_i \,;\, i \in \mathbb{Z}_{\leq m}\}; \quad k_i < k_{i+1} \ (i < m), \quad k_i = i \ (i \ll m). \tag{5.9}$$

Graphically, this sequence is determined as follows.

(5.10)

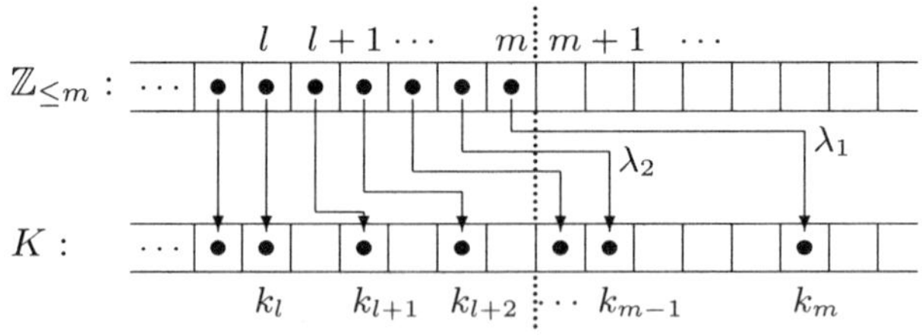

Imagine that K is the result of an event that some of the balls in the ground state $\mathbb{Z}_{\leq m}$ have moved to right. If we assume that no crossing of paths has occurred in

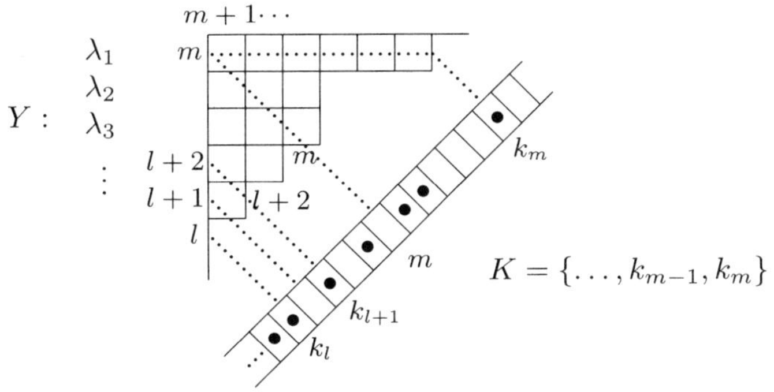

FIGURE 5.1. Maya Diagram and Young Diagram

this event, the distance of movement of each $j \in \mathbb{Z}_{\leq m}$ is given by

$$\lambda_1 = k_m - m, \quad \lambda_2 = k_{m-1} - (m-1), \quad \ldots . \tag{5.11}$$

(Note that $k_i - i$ coincides with the number of empty boxes that are located in the left of k_i. Indices for λ are set in the reversed order, in accordance with the convention of partitions.) In this way, we obtain a non-increasing sequence of integers

$$\lambda = (\lambda_1, \lambda_2, \ldots); \quad \lambda_1 \geq \lambda_2 \geq \cdots \geq 0, \quad \lambda_i = 0 \ (i \gg 0), \tag{5.12}$$

which is in fact a partition. This is the way that one can pass from the Maya diagrams K to the partitions λ (and the Young diagrams Y). We remark that the ground state $\mathbb{Z}_{\leq m}$ (of any index m) corresponds to the partition $(0, 0, \ldots)$, and hence to the empty Young diagram $\emptyset$.

PROPOSITION 5.1. *Fix an integer $m \in \mathbb{Z}$. Then the correspondence described above induces a bijection from the set of all Maya diagrams of index m to the set of all Young diagrams (partitions).*

Figure 5.1 shows how one can recover the corresponding Maya diagram (of index m) from a given Young diagram. First we put labels of integers on the left and top walls as in Figure 5.1 so that the top-left corner comes in between m and $m+1$. Then starting from the downside of the left wall, we extend the labeling along the right walls of the Young diagram. Finally we collect the labels put on the vertical walls of the Young diagram; this set K of labels gives the corresponding Maya diagram. (Each label on the right wall of Y is the same as that on the left wall joined by the dotted line as in Figure 5.1.)

Exercise: What is the Maya diagram K of index m corresponding to each of the following Young diagrams?

(5.13)

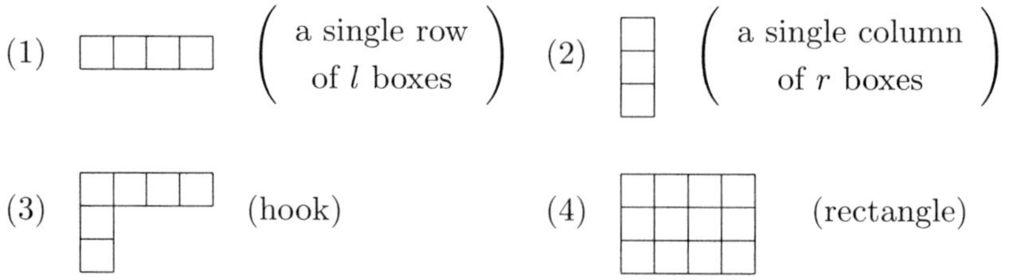

We now consider a Young diagram Y and the corresponding Maya diagram of index m. To each box s in Y, we can associate a hook having s as its corner, by stretching an arm rightward (the boxes to the right of s), and also a leg downward (the boxes below s).

(5.14)

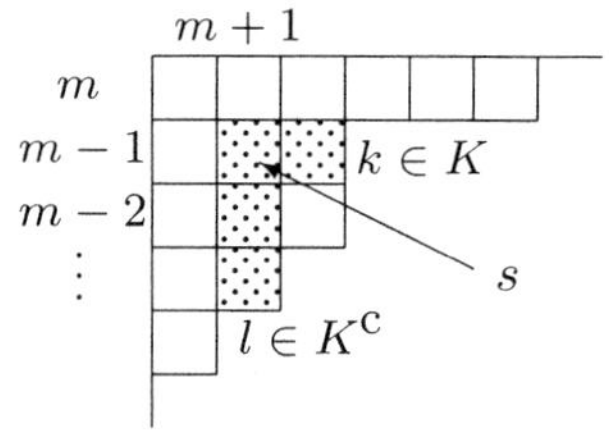

Let k be the label on the vertical wall to the right side of s, and l the label on the horizontal wall on the downward side of s. Then we have a pair (l, k) of integers such that

$$l \in K^{\mathrm{C}}, \quad k \in K, \quad l < k, \tag{5.15}$$

where we have used the notation $K^{\mathrm{C}} = \mathbb{Z} \setminus K$. In this way, we obtain a bijection between the set of boxes in Y and the set of all such pairs (l, k); the length of the hook that has s as its corner is given precisely by $k - l$.

We now explain another way of expressing partitions and Young diagrams, called the *Frobenius notation*. Consider a partition λ and its diagram Y:

(5.16)

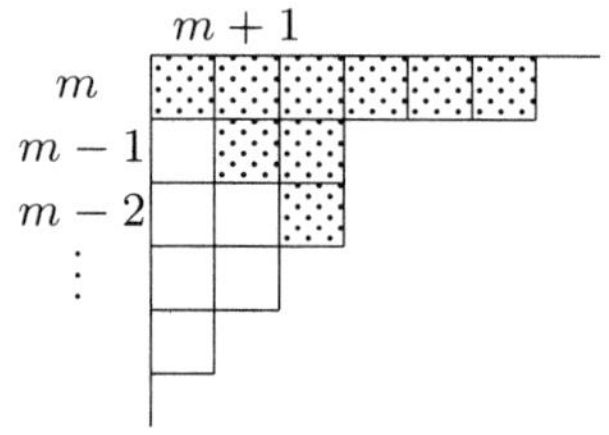

$$\lambda = (6, 3, 3, 2, 1) \\ = (6, 2, 1 | 4, 2, 0)$$

We look at the boxes in the shaded region, the upper half together with the diagonal. We denote by $a_1 > a_2 > \cdots > a_r > 0$ the lengths of rows in the shaded region, and by $b_1 > b_2 > \cdots > b_r \geq 0$ the lengths of columns in the lower half; here r stands for the length of the diagonal. Noting that the original Young diagram can be recovered from these data, we write $\lambda = (a_1, \ldots, a_r \,|\, b_1, \ldots, b_r)$, and call it the *Frobenius notation* of the partition λ. For example, when $r = 1$, $(a \,|\, b)$ represents a hook whose arm-length is $a - 1$ and whose leg-length is b. Let us consider the Maya diagram K of index m corresponding to λ. If we compare K with the ground state $\mathbb{Z}_{\leq m}$, the *excess* $I = K \setminus \mathbb{Z}_{\leq m}$ (the set of occupied boxes whose labels are $> m$)

and the *deficiency* $J = \mathbb{Z}_{\leq m} \setminus K$ (the set of empty boxes whose labels are $\leq m$) are precisely given by

$$I = \{m + a_1, \ldots, m + a_r\}, \qquad J = \{m - b_1, \ldots, m - b_r\}. \tag{5.17}$$

Remark: The definition of the Frobenius notation may depend on the literature; some authors prefer to exclude the diagonal from the a part to make a_i and b_j symmetric.

5.2. Formulas of Jacobi-Trudi type

In this section we explain what the formula of Jacobi-Trudi type is for the ϕ-factors associated with the transforms of τ-functions. Its proof will be postponed to the latter half of this book, where a detailed discussion will be given in a more general setting.

We first explain a prototype of the Jacobi-Trudi formula. Suppose that we have a rule for assigning some function s_λ to each partition $\lambda = (\lambda_1, \lambda_2, \ldots)$ (or Young diagram). We distinguish s_λ for Young diagrams of a single row from the others, and for $k = 0, 1, 2, \ldots$ we set

$$h_k = s_{(k)}; \qquad (k) = \boxed{}\boxed{}\boxed{}\boxed{\cdots}\boxed{}\boxed{} \quad (k \text{ boxes}). \tag{5.18}$$

(Young diagrams of a single row correspond to partitions $(k) = (k, 0, 0, \ldots)$ of a single part. We use the symbol h_k keeping in mind the standard notation in the theory of symmetric functions. These have nothing to do with the Hamiltonians we discussed before.) Assuming that $h_0 = 1$, for $k < 0$ we set $h_k = 0$. It happens that a general s_λ is expressed in terms of h_k ($k = 0, 1, 2, \ldots$) corresponding to one-row diagrams. We say that s_λ has the *Jacobi-Trudi formula* if it is expressed as a determinant of the form

$$\begin{aligned} s_\lambda &= \det\left(h_{\lambda_j - j + i}\right)_{i,j=1}^{\ell(\lambda)} \\ &= \det \begin{bmatrix} h_{\lambda_1} & h_{\lambda_2 - 1} & \cdots & h_{\lambda_l - l + 1} \\ h_{\lambda_1 + 1} & h_{\lambda_2} & \cdots & h_{\lambda_l - l + 2} \\ \vdots & \vdots & \ddots & \vdots \\ h_{\lambda_1 + l - 1} & h_{\lambda_2 + l - 2} & \cdots & h_{\lambda_l} \end{bmatrix} \qquad (l = \ell(\lambda)). \end{aligned} \tag{5.19}$$

Notice that the entries $h_{\lambda_1}, \ldots, h_{\lambda_l}$ on the diagonal correspond to the parts of λ; the other entries are arranged so that the indices increase one by one along each column. When $\lambda = (4, 2)$ for example, the above determinant takes the form

$$s_{(4,2)} = \det \begin{bmatrix} h_4 & h_1 \\ h_5 & h_2 \end{bmatrix} = h_4 h_2 - h_1 h_5. \tag{5.20}$$

(Note that in the Jacobi-Trudi formula for $s_{(k)}$ the right-hand side reduces to h_k, naturally.) We remark that for any choice of l such that $l \geq \ell(\lambda)$, we have $s_\lambda = \det\left(h_{\lambda_j - j + i}\right)_{1 \leq i,j \leq l}$, since $h_0 = 1$, $h_k = 0$ ($k < 0$). The original Jacobi-Trudi formula deals with the so-called *Schur functions*; the Schur function s_λ attached to any partition λ is expressed as the determinant (5.19), in terms of the *complete homogeneous symmetric functions* h_k. (As for the Schur functions, we will have a chance to discuss them in some detail in the latter half of this chapter.) In the explicit formulas for general ϕ-factors, we will encounter determinants which generalize (5.19). For this reason, in this book we will use the terminology *Jacobi-Trudi formula* to refer to

> a formula which represents some function attached to a general Young diagram as a determinant of a form similar to (5.19), in terms of functions corresponding to one-row diagrams.

We will call the determinant in (5.19) the Jacobi-Trudi formula in the original form.

The Jacobi-Trudi formula for the ϕ-factors is an extension of that for Schur functions; it is a determinant formula of the form

$$\chi_\lambda = \det\left(g^{(1-i)}_{\lambda_j-j+i}\right)_{i,j=1}^{\ell(\lambda)} \tag{5.21}$$

$$= \det\begin{bmatrix} g^{(0)}_{\lambda_1} & g^{(0)}_{\lambda_2-1} & \cdots & g^{(0)}_{\lambda_l-l+1} \\ g^{(-1)}_{\lambda_1+1} & g^{(-1)}_{\lambda_2} & \cdots & g^{(-1)}_{\lambda_l-l+2} \\ \vdots & \vdots & \ddots & \vdots \\ g^{(-l+1)}_{\lambda_1+l-1} & g^{(-l+1)}_{\lambda_2+l-2} & \cdots & g^{(-l+1)}_{\lambda_l} \end{bmatrix} \qquad (l=\ell(\lambda))$$

for a function χ_λ associated with a general partition λ. Note that the family h_k, corresponding to one-row diagrams, is replaced here by a two-parameter family of functions $g^{(r)}_k$ ($k = 0, 1, 2, \ldots;\ r \in \mathbb{Z}$) with another index r. Also, the entries of the matrix have a particular index r depending on the row[1]. When $\lambda = (4, 2)$, we have for instance

$$\chi_{(4,2)} = \det\begin{bmatrix} g^{(0)}_4 & g^{(0)}_1 \\ g^{(-1)}_5 & g^{(-1)}_2 \end{bmatrix} = g^{(0)}_4\, g^{(-1)}_2 - g^{(0)}_1\, g^{(-1)}_5. \tag{5.22}$$

In this generalization of the Jacobi-Trudi formula, it is natural to extend the family χ_λ by introducing the parameter $r \in \mathbb{Z}$ as well: For each $r \in \mathbb{Z}$, we define

$$\chi^{(r)}_\lambda = \det\left(g^{(r+1-i)}_{\lambda_j-j+i}\right)_{i,j=1}^{\ell(\lambda)}. \tag{5.23}$$

Then, for each $r \in \mathbb{Z}$ we have $g^{(r)}_k = \chi^{(r)}_{(k)}$ ($k = 0, 1, 2, \ldots$), and (5.23) certainly gives a formula expressing functions associated with general Young diagrams in terms of those associated with one-row diagrams. The original Jacobi-Trudi formulas for Schur functions are then considered as the special case of (5.23) where both $g^{(r)}_k$ and $\chi^{(r)}_\lambda$ do not depend on the parameter r.

In the case of the symmetric form of P_{IV}, as well as in the case of the discrete system of type $A^{(1)}_{n-1}$ discussed in the latter half of Chapter 4, we define the family of functions corresponding to one-row Young diagrams as follows. We first define

---

[1] In his paper entitled "Schur functions: theme and variations", I.G. Macdonald discussed several variations of the Schur functions. The Jacobi-Trudi formula we are going to discuss is his ninth variation of Schur functions. (I.G. Macdonald: Schur functions: theme and variations, Publ. I.R.M.A. Strasbourg, Actes 28e Séminaire Lotharingien, 5–39, 1992)

$g_k = g_k^{(0)}$ for each $k = 0, 1, 2, \dots$ to be the tridiagonal determinant

$$g_k = \frac{1}{N_k} \det \begin{bmatrix} f_0 & 1 & & & 0 \\ \varepsilon_{1,k} & f_1 & 1 & & \\ & \ddots & \ddots & \ddots & \\ & & \varepsilon_{k-2,k} & f_{k-2} & 1 \\ 0 & & & \varepsilon_{k-1,k} & f_{k-1} \end{bmatrix}. \tag{5.24}$$

On the diagonal, we have arranged $f_0, f_1, \dots$ by increasing the index one by one. The variables $\varepsilon_{i,j}$ $(i < j)$ on the lower subdiagonal are defined by

$$\varepsilon_{ij} = \alpha_i + \alpha_{i+1} + \cdots + \alpha_{j-1} = \varepsilon_i - \varepsilon_j \quad (i < j). \tag{5.25}$$

The denominator N_k is the normalization factor defined as the product of k terms

$$N_k = \varepsilon_{0,k}\, \varepsilon_{1,k} \cdots \varepsilon_{k-1,k}. \tag{5.26}$$

In the setting of the discrete system of type $A^{(1)}_{n-1}$, we have regarded the indices α_j, f_j as n-periodic, and have set $\alpha_0 + \cdots + \alpha_{n-1} = \delta$. For the present purpose, however, it is more convenient to extend the indexing set to $\mathbb{Z}$, rather than using the representatives $0, \dots, n-1$. The polynomials g_k for small k are given explicitly by

$$\begin{aligned} & g_0 = 1, \quad g_1 = \frac{1}{\alpha_0} f_0, \\ & g_2 = \frac{1}{(\alpha_0+\alpha_1)\alpha_1} \det \begin{bmatrix} f_0 & 1 \\ \alpha_1 & f_1 \end{bmatrix} = \frac{f_0 f_1 - \alpha_1}{(\alpha_0+\alpha_1)\alpha_1}, \\ & g_3 = \frac{1}{(\alpha_0+\alpha_1+\alpha_2)(\alpha_1+\alpha_2)\alpha_2} \det \begin{bmatrix} f_0 & 1 & 0 \\ \alpha_1+\alpha_2 & f_1 & 1 \\ 0 & \alpha_2 & f_2 \end{bmatrix} \\ & \quad = \frac{f_0 f_1 f_2 - \alpha_2 f_0 - (\alpha_1+\alpha_2) f_2}{(\alpha_0+\alpha_1+\alpha_2)(\alpha_1+\alpha_2)\alpha_2}. \end{aligned} \tag{5.27}$$

(If one uses the normalization $\alpha_0 + \alpha_1 + \alpha_2 = \delta = 1$ as in the case of P_{IV}, then the first factor of the denominator of g_3 becomes invisible.) We then define $g_k^{(r)}$ by increasing (by rotating) all the indices appearing in g_k by r:

$$g_k^{(r)} = \pi^r(g_k) \qquad (k = 0, 1, 2, \dots;\ r \in \mathbb{Z}). \tag{5.28}$$

THEOREM 5.2. *In the discrete system of type $A^{(1)}_{n-1}$ $(n \geq 3)$, define $g_k^{(r)}$ $(k = 0, 1, \dots, ; r \in \mathbb{Z})$ as above. Then for any $\nu \in L = \mathbb{Z}^n$, one can find a partition $\lambda = \lambda(\nu)$ such that the ϕ-factor ϕ_ν is expressed as the determinant*

$$\phi_\nu = N_\lambda^{(r)} \det \left(g^{(r+1-i)}_{\lambda_j - j + i} \right)_{i,j=1}^{\ell(\lambda)}, \tag{5.29}$$

where $r = \nu_1 + \cdots + \nu_n$. The normalization factor $N_\lambda^{(r)}$ is a polynomial in $\alpha_0, \alpha_1, \dots, \alpha_{n-1}$.

The proof of this theorem and its generalization will be the goal of the following chapters. In this section, we give some remarks on the statement of the theorem, and discuss some of its consequences. We will explain later in this section how to specify the partition $\lambda = \lambda(\nu)$ and the normalization factor $N_\lambda^{(r)}$. Before doing so, we give a remark on the integrality of the ϕ-factors.

THEOREM 5.3. *In the setting of Theorem* 5.2, *for each* $\nu \in L$, ϕ_ν *is a polynomial in the variables* α_j, f_j $(j = 0, 1, \ldots, n-1)$ *with coefficients in* $\mathbb{Z}$.

Proof. Recall that, for any $w \in \widetilde{W} = \langle s_0, \ldots, s_{n-1}, \pi \rangle$ (see Section 4.4), we have $w(\tau_m) = \tau_{w(\varpi_m)}$. Hence the ϕ-factor $\phi_{w,m}$ coincides with ϕ_ν for $\nu = w(\varpi_m) \in L$. In particular, $\phi_{w,m}$ is a polynomial for any $w \in \widetilde{W}$ and m. We remark that any element $w \in \widetilde{W}$ is expressed in the form $w = w_0 \pi^k$, $w_0 = s_{i_1} \cdots s_{i_p}$, $k \in \mathbb{Z}$, and that the normalization factor for $\pi^k \tau_m = \tau_{m+k}$ is 1. Hence, by expressing w_0 as a product of s_i, we can use induction on the length of the expression of w_0. Setting $\mathcal{R} = \mathbb{Z}[\alpha, f]$, we assume now that $\phi_{w,m} \in \mathcal{R}$. We expand it in the form

$$\phi_{w,m} = \sum_{\mu,\nu} c_{\mu,\nu} \alpha^\mu f^\nu \qquad (c_{\mu,\nu} \in \mathbb{Z}) \tag{5.30}$$

by using an appropriate notation of multi-indices. Then from the definition of s_i it is clear that $s_i(\phi_{w,m}) \in \mathcal{R}[f_i^{-1}]$. However, we have $\phi_{s_i w,m} = s_i(\phi_{w,m}) f_i^{k_i}$ for some integer k_i; hence we have $\phi_{s_i w,m} \in \mathcal{R}[f_i^{-1}]$. Since we already know that it is a polynomial, we must have $\phi_{s_i w,m} \in \mathcal{R}$ as desired. (Q.E.D.)

The ϕ-factors are polynomials with integer coefficients. It would be natural to expect that such polynomials will carry interesting combinatorial properties. In this respect, the ϕ-factors are sometimes called the *special polynomials* associated with τ-functions.

We now explain how to specify the normalization factor $N_\lambda^{(r)}$ that appeared in Theorem 5.2. It is defined as follows by using $r \in \mathbb{Z}$ and the partition λ. We take the Maya diagram $K \subset \mathbb{Z}$ of index r corresponding to the partition λ, as explained in the previous section. Then the normalization factor is given by

$$N_\lambda^{(r)} = \prod_{i \in K^{\mathrm{C}}, j \in K;\ i<j} \varepsilon_{i,j}. \tag{5.31}$$

Here the product is taken over all pairs (i, j) corresponding to the boxes in the Young diagram. Let us show an example of the case where $r = 2$ and $\lambda = (5, 3, 1)$. In the Young diagram Y corresponding to λ, we first put a number in each box as follows.

(5.32)

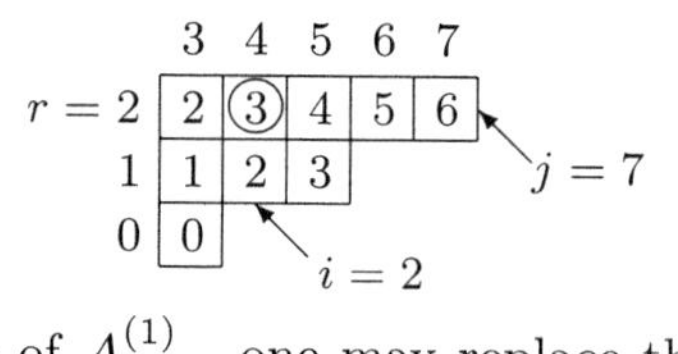

$$\varepsilon_{2,7} = \alpha_2 + \alpha_3 + \alpha_4 + \alpha_5 + \alpha_6 = \varepsilon_2 - \varepsilon_7$$

(In the case of $A_{n-1}^{(1)}$, one may replace the numbers by $0, 1, \ldots, n-1$ taking representatives in $\mathbb{Z}/n\mathbb{Z}$.) In this example, we have $K = \{\ldots, -2, -1, 1, 4, 7\}$. Let us look at the box s corresponding to the pair $(i, j) = (2, 7)$, which is now marked by ③. Note that the hook corresponding to s has length 5. The factor ε_{ij} is then computed as the sum of α_k over all numbers k picked up from the boxes belonging to the hook. (Note that, when $\alpha_k = 1$ for all k, ε_{ij} coincides with the hook length.) In this sense, $\varepsilon_{i,j}$ can be regarded as the *α-weighted length* of the hook. The normalization factor $N_\lambda^{(r)}$ is then determined as the product of $|Y|$ factors representing the α-weighted hook lengths. When $r = 0$, and $\lambda = (k)$ is a single row of length k, $N_\lambda^{(r)}$ reduces to the normalization factor

$$N_{(k)}^{(0)} = \varepsilon_{0,k}\,\varepsilon_{1,k}\,\cdots\,\varepsilon_{k-1,k} = N_k \tag{5.33}$$

for g_k.

Concerning Theorem 5.2, it remains to explain how one should specify the partition $\lambda = \lambda(\nu)$ for each $\nu = (\nu_1, \dots, \nu_n) \in L = \mathbb{Z}^n$. When an element $\nu \in L$ is given, we define the subset K of $\mathbb{Z}$ to be the union

$$K = (n\mathbb{Z}_{<\nu_1} + 1) \cup (n\mathbb{Z}_{<\nu_2} + 2) \cup \cdots \cup (n\mathbb{Z}_{<\nu_n} + n)\,. \tag{5.34}$$

Then K becomes a Maya diagram of index $r = \nu_1 + \cdots + \nu_n$. The partition $\lambda = \lambda(\nu)$ is then defined to be the one that corresponds to K in the way explained in the previous section. We now explain the meaning of this construction by using the example $\nu = (1, 2, -1)$ for $n = 3$. We first take the three ground states having indices ν_1, ν_2, ν_3 respectively, and pile them up as follows:

(5.35)

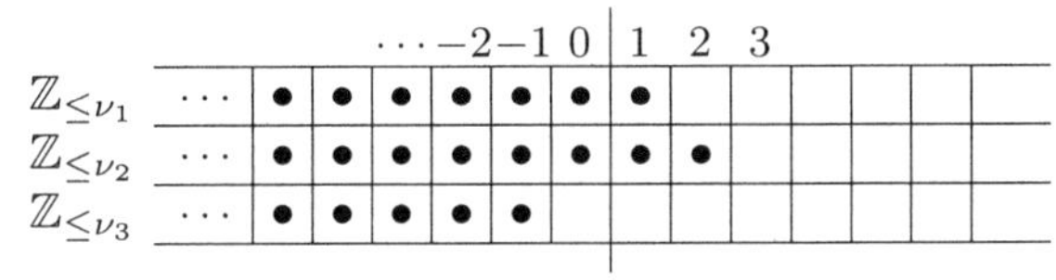

By using the numbering

(5.36)

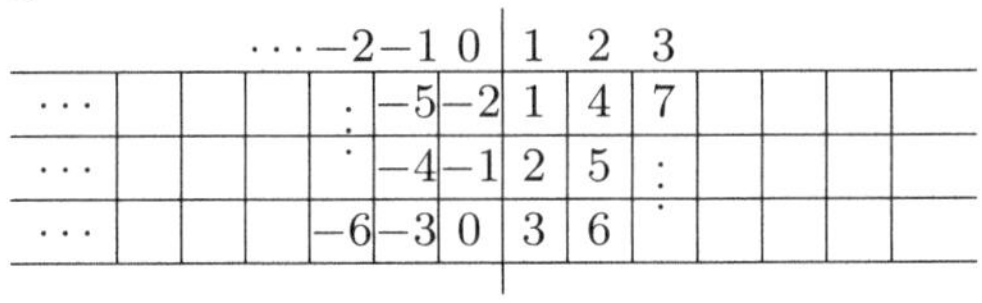

of the board, we transform the bunch of Maya diagrams above into a single Maya diagram. The result is the Maya diagram $K = \{\dots, -2, -1, 1, 2, 5\}$ of index $r = 2$:

(5.37)

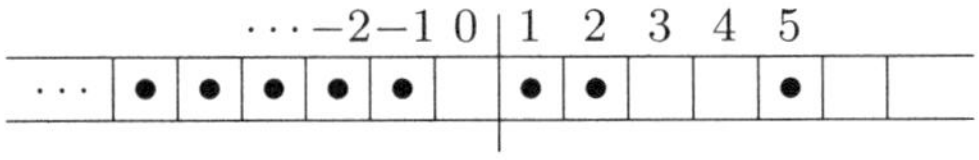

Thus the partition $\lambda = \lambda(\nu)$ is determined as $\lambda = (3, 1, 1)$.

Note that the partition λ remains unchanged in this construction if one replaces $\nu = (\nu_1, \dots, \nu_n)$ by its shift $\mu = (\nu_1 + k, \dots, \nu_n + k)$ $(k \in \mathbb{Z})$. Although the index $r = |\nu|$ may vary, the difference $|\mu| - |\nu|$ is a multiple of n; hence we have the same Jacobi-Trudi formula. This is in accordance with the fact that $\phi_\mu = \phi_\nu$. It should be noted that not all the Young diagrams appear in this construction. Such Young diagrams that appear in this way are precisely those which have no hook whose length is a multiple of n (such a Young diagram will be called an *n-core*). This point will be clarified in the course of the proof of the Jacobi-Trudi formula.

We remark that both the definition of $g_k^{(r)}$ and the right-hand side of our Jacobi-Trudi formula are independent of the integer n appearing in the Weyl group type

$A^{(1)}_{n-1}$. This means that they can be thought of as functions in infinitely many variables α_j and f_j ($j \in \mathbb{Z}$). Furthermore, we have no strong reason to restrict the definition $g^{(r)}_k$ to tridiagonal matrices. In view of this fact, we introduce the set of infinitely many variables

$$\alpha_j \quad (j \in \mathbb{Z}), \qquad f_{ij} \quad (i, j \in \mathbb{Z};\ i < j) \tag{5.38}$$

and define $f_i = f_{i\,i+1}$ ($i \in \mathbb{Z}$). (When we use the ε variables ε_j ($j \in \mathbb{Z}$), we set $\alpha_j = \varepsilon_j - \varepsilon_{j+1}$.) In this general setting, we define

$$g_k = \frac{1}{N_k} \det \begin{bmatrix} f_0 & f_{02} & f_{03} & \cdots & f_{0,k} \\ \varepsilon_{1,k} & f_1 & f_{13} & & f_{1,k} \\ & \ddots & \ddots & \ddots & \vdots \\ & & \varepsilon_{k-2,k} & f_{k-2} & f_{k-2,k} \\ 0 & & & \varepsilon_{k-1,k} & f_{k-1} \end{bmatrix} \tag{5.39}$$

and apply the transformation π such that

$$\pi(\alpha_j) = \alpha_{j+1}, \qquad \pi(f_{ij}) = f_{i+1,j+1} \tag{5.40}$$

to define $g^{(r)}_k = \pi^r(g_k)$ ($r \in \mathbb{Z}$). Then, for an arbitrary partition λ we can consider the Jacobi-Trudi determinant

$$\Phi_\lambda = N_\lambda \det \left(g^{(1-i)}_{\lambda_j - j + i} \right)_{i,j=1}^{\ell(\lambda)}, \qquad \Phi^{(r)}_\lambda = \pi^r(\Phi_\lambda). \tag{5.41}$$

As we will see later in Chapter 7, these polynomials in infinitely many variables are nothing but the ϕ-factors associated with the realization of the *Weyl group of type* A_∞ as a group of birational transformations. It also turns out that Φ_λ are in fact polynomials in α_j, f_{ij} ($i, j \in \mathbb{Z};\ i < j$) with coefficients in $\mathbb{Z}$. The ϕ-factors of the Weyl group type $A^{(1)}_{n-1}$ are then obtained by specializing the variables α_j, f_{ij}.

5.3. Examples of $P_{\rm IV}$ and $P_{\rm II}$

Let us consider the case of $P_{\rm IV}$ of the Weyl group of type $A^{(1)}_2$. In the following, we use the normalization $\delta = 1$, and keep the notation of Sections 4.2 and 4.3. Figure 5.2 indicates the partitions λ corresponding to the ϕ-factors $\phi_{m,n} = \phi_{m,n,0}$ for small m, n. Recall that we have seen the example $\lambda(1, 2, -1) = (3, 1, 1)$ in the previous section. Since $(1, 2, -1) \equiv (2, 3, 0)$, we also have $\lambda(2, 3, 0) = (3, 1, 1)$. Please check that the partition corresponding to $\phi_{2,3}$ is shown to be $(3, 1, 1)$ in Figure 5.2.

Let us look at simpler examples:

$$\begin{aligned} \phi_{2,1} &= G_1 &&\Rightarrow &\lambda &= (1), & r &= 3, \\ \phi_{2,0} &= F_2 &&\Rightarrow &\lambda &= (2), & r &= 2. \end{aligned} \tag{5.42}$$

In these cases, the normalization factors cancel out, so that

$$G_1 = f_0, \quad F_2 = \det \begin{bmatrix} f_2 & 1 \\ \alpha_0 & f_0 \end{bmatrix} = f_2 f_0 - \alpha_0. \tag{5.43}$$

The first nontrivial Jacobi-Trudi formula is the case of $\phi_{2,2}$:

$$\phi_{2,2} \ \Rightarrow\ \lambda = (1, 1), \quad r = 4 \equiv 1. \tag{5.44}$$

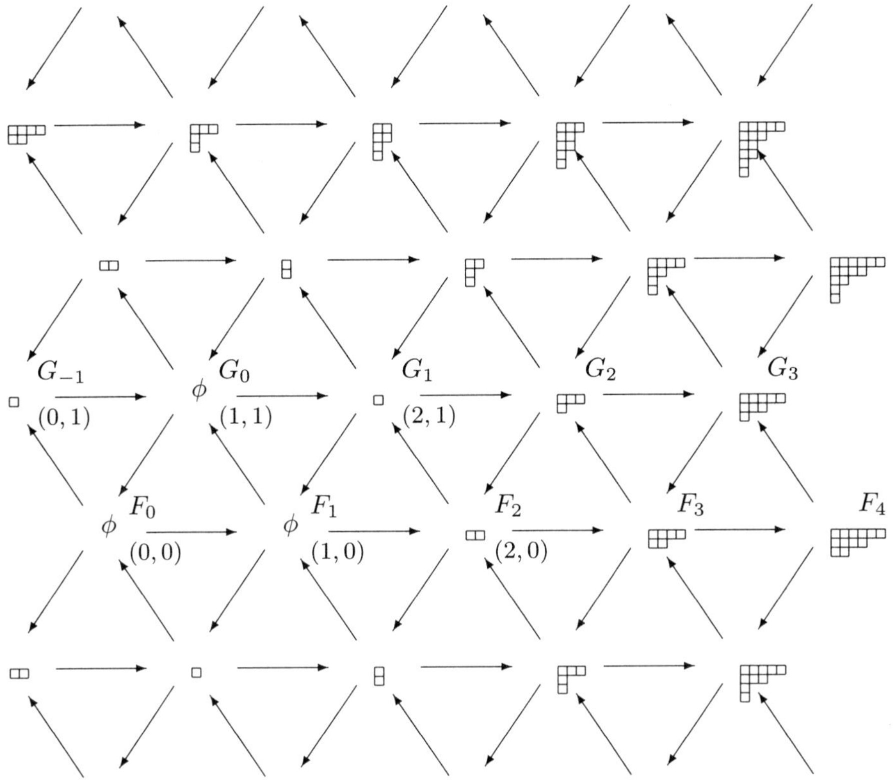

FIGURE 5.2. Young Diagrams for Determining $\phi_{m,n}$

In this case we have

$$(5.45)\quad \begin{aligned} \frac{\phi_{2,2}}{N^{(1)}_{(1,1)}} &= \det \begin{bmatrix} g_1^{(1)} & 1 \\ g_2^{(0)} & g_1^{(0)} \end{bmatrix} = g_1^{(1)} g_1^{(0)} - g_2^{(0)} \\ &= \frac{f_1 f_0}{\alpha_1 \alpha_0} - \frac{f_0 f_1 - \alpha_1}{(\alpha_0+\alpha_1)\alpha_1} = \frac{f_0 f_1 + \alpha_0}{(\alpha_0+\alpha_1)\alpha_0}. \end{aligned}$$

On the other hand, we have $N^{(1)}_{(1,1)} = (\alpha_0+\alpha_1)\alpha_0$, and hence

$$(5.46)\quad \phi_{2,2} = f_0 f_1 + \alpha_0.$$

Exercise: Compute $\phi_{3,1} = G_2$ in the same way to prove

$$(5.47)\quad \phi_{3,1} = G_2 = f_0^2 f_1 f_2 - \alpha_0 f_0 f_1 - (\alpha_0+\alpha_2) f_0^2 + \alpha_0 f_0 f_2 - \alpha_0^2.$$

Exercise: Show that the partitions associated with $F_m = \phi_{m,0,0}$ and $G_m = \phi_{m+1,1,0}$ for $m \geq 0$ are

$$(5.48)\quad \begin{aligned} \lambda(m,0,0) &= (2m-2, 2m-4, \ldots, 2), \\ \lambda(m+1,1,0) &= (2m-1, 2m-3, \ldots, 1). \end{aligned}$$

What about the case when $m < 0$?

In Theorem 5.2 we have excluded the case of $A_1^{(1)}$. We remark however that the Jacobi-Trudi formula for this case of $P_{\rm II}$ is obtained simply by modifying the definition of g_k. This case can also be treated in the framework of the ϕ-factors Φ_λ in infinitely many variables that was introduced at the end of the previous section.

In the discussion of $P_{\rm II}$, we considered the shift operator $T = \pi s_1$ in the group $\widetilde{W} = \langle\, s_0, s_1, \pi \,\rangle$. In order to make the notation consistent with the case of $P_{\rm IV}$, we could use the two shift operators

$$T_1 = \pi s_1, \quad T_2 = s_1 \pi, \tag{5.49}$$

so that $T_1 T_2 = T_2 T_1 = 1$. (Since $T_2 = T^{-1}$, we did not need to use T_2 explicitly.) Formally, we could also define

$$\tau_{m,n} = T_1^m T_2^n(\tau_0) = \phi_{m,n}\tau_0 \left(\frac{\tau_1}{\tau_0}\right)^m \left(\frac{\tau_0}{\tau_1}\right)^n = \phi_{m,n}\tau_0^{1-m+n}\tau_1^{m-n} \tag{5.50}$$

for $m, n \in \mathbb{Z}$. But, as for the ϕ-factors, we need only consider $\phi_n = \phi_{n,0}$, since $\phi_{m,n} = \phi_{m-n,0}$.

For the Jacobi-Trudi formula of this case, we define $g_k = g_k^{(0)}$ as follows:

$$g_k = \frac{1}{N_k} \det \begin{bmatrix} f_0 & -2q & -2 & & & 0 \\ \varepsilon_{1,k} & f_1 & 2q & -2 & & \\ & \ddots & \ddots & \ddots & \ddots & \\ & & & & & -2 \\ & & & \varepsilon_{k-2,k} & f_{k-2} & (-1)^{k-1}2q \\ 0 & & & & \varepsilon_{k-1,k} & f_{k-1} \end{bmatrix}. \tag{5.51}$$

Except for the fact that α_j and f_j are 2-periodic with respect to the indices, the diagonal and the lower subdiagonal parts have the same structure as before. In this case, we use the 4-diagonal determinant; in the upper subdiagonal, we arrange $-2q$ and $2q$ one after the other; we also put -2's on top of them. Note that we have two types of $g_k^{(r)} = \pi^r(g_k)$ according to the parity of $r \in \mathbb{Z}$. Once we have defined g_k as above, the rest of the construction of the Jacobi-Trudi formula is the same as before.

When $n \geq 0$, the Maya diagram K for $\phi_n = \phi_{n,0}$ is given by

$$K = (2\mathbb{Z}_{<n} + 1) \cup (2\mathbb{Z}_{<0} + 2) = \{\ldots, -1, 0, 1, 3, \ldots, 2n-1\}. \tag{5.52}$$

Note that K has index n. Hence we obtain the partition $\lambda = (n-1, n-2, \ldots, 1)$; the corresponding Young diagram is the *staircase* of $n-1$ steps. For $n = 0, 1$, we have $Y = \emptyset$ (the empty set). For $n = 2, 3, 4, 5, \ldots$, we can write down the Jacobi-Trudi formula by using the Young diagrams labeled as follows:

$$\begin{array}{|c|}\hline 0\\\hline\end{array} \qquad \begin{array}{|c|c|}\hline 1&0\\\hline 0\\\cline{1-1}\end{array} \qquad \begin{array}{|c|c|c|}\hline 0&1&0\\\hline 1&0\\\cline{1-2} 0\\\cline{1-1}\end{array} \qquad \begin{array}{|c|c|c|c|}\hline 1&0&1&0\\\hline 0&1&0\\\cline{1-3} 1&0\\\cline{1-2} 0\\\cline{1-1}\end{array} \qquad \cdots \tag{5.53}$$

Note that the number at the top-left corner represents $r \in \mathbb{Z}/2\mathbb{Z}$.

5.4. Schur functions and ϕ-factors

The original Jacobi-Trudi formulas are the determinant expressions for *Schur functions*. In this section, we recall the Jacobi-Trudi formulas for the Schur functions, and discuss the relationship with the ϕ-factors for the discrete system of type $A^{(1)}_{n-1}$. The Schur functions are the *real* special polynomials; they have been attracting a wide range of mathematicians (including the author). There are a great number of topics concerning the Schur functions, but we shall take a modest approach to them here. According to a standard definition, the Schur functions are a family of symmetric polynomials $s_\lambda(x)$ in N variables $x = (x_1, \ldots, x_N)$, indexed by the partitions λ with $\ell(\lambda) \le N$. They provide a basis for the ring of symmetric polynomials; any symmetric polynomial in x is expressed uniquely as a linear combination of the Schur functions $s_\lambda(x)$ ($\ell(\lambda) \le N$). In the following, however, we introduce the Schur functions as they appear in the context of the KP hierarchy[2]. (We refer later to the relationship with the standard definition mentioned above.)

We consider a set of infinitely many variables $\boldsymbol{t} = (t_1, t_2, \ldots)$, and define a sequence of polynomials $p_l(\boldsymbol{t}) = p_l(t_1, t_2, \ldots)$ ($l = 0, 1, \ldots$) in these variables through the generating function

$$\sum_{l=0}^{\infty} p_l(\boldsymbol{t}) u^l = \exp(\eta(\boldsymbol{t}, u)), \qquad \eta(\boldsymbol{t}, u) = \sum_{k=1}^{\infty} t_k u^k. \tag{5.54}$$

Although the number of variables is infinite, by a polynomial in $\boldsymbol{t}$, we mean a polynomial in some finite number of variables amongst $t_1, t_2, \ldots$; the finite set of variables may vary depending on the polynomial. When we talk about the degree of a polynomial in $\boldsymbol{t}$, we count its degree with weights by assigning a degree i to each t_i. For each $l = 0, 1, \ldots$, the polynomial $p_l(\boldsymbol{t})$ is homogeneous of degree l in the sense mentioned above. (This is because the variable u in the generating function keeps counting the degree.) The precise meaning of the definition of $p_l(\boldsymbol{t})$ is the following: Expand the right-hand side as a power series in u, and define $p_l(\boldsymbol{t})$ to be the coefficient of u^l. Let us try:

$$\exp(\eta(\boldsymbol{t}, u)) = \sum_{n=0}^{\infty} \frac{1}{n!} \eta(\boldsymbol{t}, u)^n = \sum_{n=0}^{\infty} \frac{1}{n!} \left(t_1 u + t_2 u^2 + \cdots\right)^n. \tag{5.55}$$

By the multinomial expansion we have

$$\begin{aligned} \frac{1}{n!}\left(t_1 u + t_2 u^2 + \cdots\right)^n &= \sum_{k_1 + k_2 + \cdots = n} u^{k_1 + 2k_2 + \cdots} \frac{t_1^{k_1}}{k_1!} \frac{t_1^{k_2}}{k_2!} \cdots \\ &= \sum_{l=0}^{\infty} u^l \sum_{|k| = n, \|k\| = l} \frac{t_1^{k_1}}{k_1!} \frac{t_1^{k_2}}{k_2!} \cdots . \end{aligned} \tag{5.56}$$

Here we have used the notations $|k| = \sum_i k_i$ and $\|k\| = \sum_i i k_i$ for each $k = (k_1, k_2, \ldots)$ such that $k_i = 0$ for $i \gg 0$. We substitute these into the above expansion of the exponential function, and then take the coefficient of u^l. As the result we obtain the expression

$$\begin{aligned} p_l(\boldsymbol{t}) &= \sum_{\|k\| = l} \frac{t_1^{k_1}}{k_1!} \frac{t_1^{k_2}}{k_2!} \cdots \\ &= \frac{t_1^l}{l!} + \frac{t_1^{l-2} t_2}{(l-2)!} + \cdots + t_{l-1} t_1 + t_l. \end{aligned} \tag{5.57}$$

---

[2] T. Miwa, M. Jimbo and E. Date, *Solitons : Differential Equations, Symmetries and Infinite-Dimensional Algebras*, Cambridge Tracts in Mathematics, 135, Cambridge University Press, 2000.

If you feel uneasy doing such a formal computation, please try picking up the expansion coefficients of u^l explicitly for small l. To each $k = (k_1, k_2, \ldots)$, one can associate a partition consisting of k_1 1's, k_2 2's and so on, so that $k = (k_1, k_2, \ldots)$ with $\|k\| = l$ corresponds bijectively to a partition of l. Hence the number of monomials appearing in the expression of $p_l(\boldsymbol{t})$ above is precisely the partition number of l.

We now take an arbitrary partition $\lambda = (\lambda_1, \lambda_2, \ldots)$, and define the polynomial $S_\lambda = S_\lambda(\boldsymbol{t})$ by the determinant

$$
\begin{aligned}
S_\lambda &= \det \left(p_{\lambda_j - j + i}\right)_{i,j=1}^{\ell(\lambda)} \\
&= \det \begin{bmatrix} p_{\lambda_1} & p_{\lambda_2 - 1} & \cdots & p_{\lambda_l - l + 1} \\ p_{\lambda_1 + 1} & p_{\lambda_2} & \cdots & p_{\lambda_l - l + 2} \\ \vdots & \vdots & \ddots & \vdots \\ p_{\lambda_1 + l - 1} & p_{\lambda_2 + l - 2} & \cdots & p_{\lambda_l} \end{bmatrix} \qquad (l = \ell(\lambda)).
\end{aligned} \tag{5.58}
$$

Here we have also suppressed the variables $\boldsymbol{t}$ for $p_l = p_l(\boldsymbol{t})$. By definition, $S_\lambda(\boldsymbol{t})$ is a homogeneous polynomial in $\boldsymbol{t}$ of degree $|\lambda|$ (with weights such that $\deg t_i = i$). Since we have defined S_λ by means of the Jacobi-Trudi determinant, it is tautological to say that S_λ *has the form of a Jacobi-Trudi formula.* But this will make sense if we clarify the relation between $S_\lambda(\boldsymbol{t})$ and the Schur function $s_\lambda(x)$ in N variables $x = (x_1, \ldots, x_N)$. In fact, if we substitute $\frac{1}{k}$ of the *power sum* of $x = (x_1, \ldots, x_N)$

$$
t_k(x) = \frac{1}{k}\left(x_1^k + x_2^k + \cdots + x_N^k\right) \qquad (k = 1, 2, \ldots) \tag{5.59}
$$

into each t_k $(k = 1, 2, \ldots)$, then the polynomial S_λ gives rise to the Schur function in the N variables; namely, we have $s_\lambda(x) = S(\boldsymbol{t}(x))$. In this sense, we have Jacobi-Trudi formulas for the Schur functions.

Usually, the Schur function $s_\lambda(x)$ in N variables $x = (x_1, \ldots, x_N)$ is defined as the ratio

$$
s_\lambda(x) = \frac{\det\left(x_i^{\lambda_j + N - j}\right)_{i,j=1}^N}{\det\left(x_i^{N-j}\right)_{i,j=1}^N} \tag{5.60}
$$

of two determinants. The denominator is the Vandermonde determinant, and hence is equal to the difference product $\prod_{1 \le i < j \le N}(x_i - x_j)$. Since the numerator is an alternating polynomial, it is divisible by the difference product. Hence, $s_\lambda(x)$ is a symmetric polynomial after all. In general, it is known that the Schur function $s_\lambda(x)$ is the generating function for the so-called Young tableaux (obtained by putting numbers into the boxes of the Young diagram of the partition λ, under certain rules). In particular, it is expressed as a linear combination of monomials with positive integer coefficients. In the case of a single row, for example, we have

$$
\begin{aligned}
s_{(l)} = h_l(x) &= \sum_{k_1 + \ldots + k_N = l} x_1^{k_1} \cdots x_N^{k_N} \\
&= \sum_{1 \le j_1 \le \cdots \le j_l \le N} x_{j_1} \cdots x_{j_l} \qquad (l = 0, 1, \ldots).
\end{aligned} \tag{5.61}
$$

This $h_l(x)$ is called the *complete homogeneous symmetric function* of degree l. In the case of a single column,

$$s_{(1^r)} = e_r(x) = \sum_{1\le j_1<\cdots<j_r\le N} x_{j_1}\cdots x_{j_r} \quad (r=0,1,\ldots,N) \tag{5.62}$$

is the *elementary symmetric function* of degree r, where we have used the abbreviation $(1^r) = (1,\ldots,1)$ (r times). The complete homogeneous symmetric functions and the elementary symmetric functions have the following generating functions, respectively:

$$\begin{aligned} \sum_{l=0}^{\infty} h_l(x)\, u^l &= \frac{1}{(1-x_1u)\cdots(1-x_Nu)}, \\ \sum_{r=0}^{N} e_r(x)\, u^r &= (1+x_1u)\cdots(1+x_Nu). \end{aligned} \tag{5.63}$$

Exercise: Prove that the following relations hold under the specialization (5.59):

$$p_l(\boldsymbol{t}(x)) = h_l(x), \quad (-1)^r p_r(-\boldsymbol{t}(x)) = e_r(x). \tag{5.64}$$

(Compare the generating function for p_l with those for h_l and e_r given above.)

In what follows, we will mainly use the name *Schur functions* to refer to $S_\lambda(\boldsymbol{t})$ as in the context of the KP hierarchy.

The Schur functions $S_\lambda(\boldsymbol{t})$ are closely related to our ϕ-factors. A reason for this is that each $p_k = p_k(\boldsymbol{t})$ is expressed by a determinant in terms of $t_1, t_2, \ldots$.

LEMMA 5.4. *For each $k = 0, 1, \ldots$, we have*

$$p_k(\boldsymbol{t}) = \frac{1}{k!} \det \begin{bmatrix} t_1 & 2t_2 & 3t_3 & \cdots & kt_k \\ -k+1 & t_1 & 2t_2 & & \\ & -k+2 & t_1 & \ddots & \vdots \\ & & \ddots & \ddots & 2t_2 \\ 0 & & & -1 & t_1 \end{bmatrix}. \tag{5.65}$$

This lemma can be proved as follows. We denote by $\Phi(u)$ the left-hand side of the definition (5.54), regarding it as a function (power series) of u. By applying the differential operator $u\frac{d}{du}$ to (5.54), we obtain the differential equation

$$u\frac{d\Phi}{du} = u\frac{d\eta(\boldsymbol{t},u)}{du}\Phi(u), \quad u\frac{d\eta(\boldsymbol{t},u)}{du} = \sum_{i=1}^{\infty} it_i u^i \tag{5.66}$$

for $\Phi(u)$. If we pick up the coefficients of u^k ($k = 1, 2, \ldots$), we can rewrite the above formula into a system of an infinite number of linear equations

$$kp_k = \sum_{i+j=k} i\,t_i\,p_j, \quad \text{i.e.} \quad -kp_k + \sum_{i=1}^{k-1} i\,t_i\,p_{k-i} = -kt_k. \tag{5.67}$$

By means of matrices, this system is expressed equivalently as

$$\begin{bmatrix} -k & t_1 & 2t_2 & \cdots & (k-1)t_{k-1} \\ & \ddots & \ddots & & \vdots \\ & & -3 & t_1 & 2t_2 \\ & & & -2 & t_1 \\ 0 & & & & -1 \end{bmatrix} \begin{bmatrix} p_k \\ \vdots \\ p_3 \\ p_2 \\ p_1 \end{bmatrix} = - \begin{bmatrix} kt_k \\ \vdots \\ 3t_2 \\ 2t_2 \\ t_1 \end{bmatrix}. \tag{5.68}$$

Using the Cramer formula, we can solve this system of linear equations, to obtain the determinant formula for p_k as presented above.

There is an obvious analogy between (5.65) of this lemma and the definition (5.24) of g_k, or (5.39) in general, used in the Jacobi-Trudi formula for the ϕ-factors. In the setting of ϕ-factors in infinitely many variables, the $p_k(\boldsymbol{t})$ are recovered essentially by setting

$$\alpha_j = -1, \qquad f_{j,j+k} = kt_k \quad (k = 1, 2, \ldots) \tag{5.69}$$

in (5.39). Under this specialization, we have in fact

$$g_k^{(r)} = (-1)^k p_k(\boldsymbol{t}) \qquad (k = 0, 1, \ldots) \tag{5.70}$$

for any $r \in \mathbb{Z}$. The normalization factor N_λ coincides, up to sign, with the product of all hook lengths, which we denote by H_λ, of the Young diagram corresponding to λ: $N_\lambda = (-1)^{|\lambda|} H_\lambda$. In this way, we obtain

$$\Phi_\lambda = H_\lambda \, S_\lambda(\boldsymbol{t}), \tag{5.71}$$

where the minus signs have canceled out. (Note also that H_λ is a specialization of N_λ such that $\alpha_j = 1$ for all j.) Another way of specialization is

$$\alpha_j = 1, \qquad f_{j,j+k} = (-1)^{k-1} kt_k \quad (k = 1, 2, \ldots). \tag{5.72}$$

Under this specialization, we have $g_k^{(r)} = p_k(\boldsymbol{t})$ $(k = 0, 1, \ldots)$ for any $r \in \mathbb{Z}$. (Note that the determinant remains invariant even after replacing the (i, j)-component by $(-1)^{i+j}$.) In this case again, we have the same formula $\Phi_\lambda = H_\lambda S_\lambda(\boldsymbol{t})$.

In the case of the discrete system of type $A_{n-1}^{(1)}$ (Section 4.4), we come to the setting of ϕ-factors by specializing $\boldsymbol{t}$ further to

$$\alpha_j = \pm 1; \quad t_1 = t, \quad t_2 = \mp \frac{1}{2}, \quad t_i = 0 \quad (i \geq 3). \tag{5.73}$$

Hence we have

THEOREM 5.5. *In the discrete system of type* $A_{n-1}^{(1)}$ $(n \geq 3)$ *with* $\delta = \pm n$, *under the specialization*

$$\alpha_j = \pm 1, \quad f_j = t \qquad (j = 0, 1, \ldots, n-1), \tag{5.74}$$

the ϕ*-factor for any* $\nu \in L = \mathbb{Z}^n$ *is represented by the Schur function attached to the* n*-core* $\lambda = \lambda(\nu)$ *as follows*:

$$\phi_\nu = H_\lambda \, S_\lambda(\boldsymbol{t}), \quad \boldsymbol{t} = \left(t, \mp\frac{1}{2}, 0, \ldots\right). \tag{5.75}$$

In the language of τ-functions on the lattice, the theorem above implies that there is a particular solution

$$\tau_\nu = H_\lambda\, S_\lambda(\boldsymbol{t})\, \tau_0^{1-\nu_1+\nu_n} \tau_1^{\nu_1-\nu_2} \cdots \tau_{n-1}^{\nu_{n-1}-\nu_n} \qquad (\nu \in L) \tag{5.76}$$

with initial values $\tau_0, \dots, \tau_{n-1} \in \mathbb{C}$. If we take α_j, f_j specified above as the initial values, we see that, for the corresponding solution of the discrete system, each $T^\nu(f_j)$ is expressed in terms of four Schur functions. We remark that the Okamoto polynomials $Q_n(x)$ in Table 4.1 correspond to the case where $n = 3$ and $\delta = -3$, so that

$$\begin{aligned} &Q_n(x) = H_\lambda S_\lambda(x, \tfrac{1}{2}, 0, \dots), \\ &\lambda = \begin{cases} (2n-2, 2n-3, \dots, 2) & (n > 0), \\ (-n, -n, \dots, 2, 2, 1, 1) & (n \le 0). \end{cases} \end{aligned} \tag{5.77}$$

(The polynomials $R_n(x)$ also have similar expressions including other Young diagrams.)

The situation is essentially the same with the Okamoto polynomials F_n, G_n for P_{IV} and the Yablonski-Vorobiev polynomials for P_{II} which were discussed in Chapter 4, although a modification should be made according to the normalization $\delta = 1$. Omitting the details of this modification, we only formulate the results.

THEOREM 5.6. *In the case of the symmetric form of P_{IV}, the ϕ-factors ϕ_ν $(\nu \in \mathbb{Z}^3)$ corresponding to the rational solution $\alpha_j = 1/3$, $f_j = t/3$ $(j = 0, 1, 2)$ are expressed as*

$$\phi_\nu = 3^{-|\lambda|} H_\lambda\, S_\lambda \left(t, -\frac{3}{2}, 0, \dots\right) \tag{5.78}$$

by means of the Schur functions associated with the 3-cores $\lambda = \lambda(\nu)$.

Similarly we have

THEOREM 5.7. *In the case of the symmetric form of P_{II}, the ϕ-factors ϕ_n $(n \in \mathbb{Z}, n > 0)$ corresponding to the rational solution $\alpha_j = 1/2$, $f_j = t/2$ $(j = 0, 1)$, $q = 0$ are expressed as*

$$\phi_n = 2^{-n(n-1)/2} H_\lambda\, S_\lambda \left(t, 0, -\frac{4}{3}, 0, \dots\right) \tag{5.79}$$

by means of the Schur functions associated with the staircases $\lambda = (n-1, n-2, \dots, 1)$.

Accordingly, the Yablonski-Vorobiev polynomials P_n $(n > 0)$ are expressed as

$$P_n(t) = H_\lambda\, S_\lambda \left(t, 0, -\frac{4}{3}, 0, \dots\right) \tag{5.80}$$

by using the staircases of $(n-1)$ steps. (Note that $P_n = P_{1-n}$ for $n \le 0$.) We finally remark that the coefficient of $t_1^{|\lambda|}$ in $S_\lambda(\boldsymbol{t})$ coincides with the reciprocal of H_λ, for arbitrary λ. Hence, the combination $H_\lambda\, S_\lambda(\boldsymbol{t})$ has the effect of normalizing the coefficient of $t_1^{|\lambda|}$ (of the highest degree) to be 1.

CHAPTER 6

Getting Familiar with Determinants

In this chapter we deal with a slightly different subject — matrices and determinants. We discuss the Gauss decomposition of matrices and related topics, such as the Jacobi identity and the Plücker relations. Some of the explicit formulas we are going to introduce below will be used later in studying the structure of rational transformations arising from Bäcklund transformations of the Painlevé equations.

6.1. Basic properties of minor determinants

Let $X = (x_{ij})_{1\le i\le m, 1\le j\le n}$ be an $m \times n$ matrix:

$$X = \begin{bmatrix} x_{11} & x_{12} & \cdots & x_{1n} \\ x_{21} & x_{22} & \cdots & x_{2n} \\ \vdots & \vdots & \ddots & \vdots \\ x_{m1} & x_{m2} & \cdots & x_{mn} \end{bmatrix}. \tag{6.1}$$

We denote by

$$X^{i_1\dots i_r}_{j_1\dots j_r} = \begin{bmatrix} x_{i_1j_1} & x_{i_1j_2} & \cdots & x_{i_1j_r} \\ x_{i_2j_1} & x_{i_2j_2} & \cdots & x_{i_2j_n} \\ \vdots & \vdots & \ddots & \vdots \\ x_{i_rj_1} & x_{i_rj_2} & \cdots & x_{i_rj_r} \end{bmatrix} \tag{6.2}$$

the $r \times r$ submatrix of X with row indices $i_1, \dots, i_r$ and column indices $j_1, \dots, j_r$; its determinant

$$\xi^{i_1\dots i_r}_{j_1\dots j_r}(X) = \det(X^{i_1\dots i_r}_{j_1\dots j_r}) = \sum_{\sigma\in\mathfrak{S}_r} \epsilon(\sigma) x_{i_{\sigma(1)}j_1} x_{i_{\sigma(2)}j_2} \cdots x_{i_{\sigma(r)}j_r} \tag{6.3}$$

is called the *minor determinant* of degree r with respect to the rows $i_1, \dots, i_r$ and the columns $j_1, \dots, j_r$. Hereafter, we denote by $\mathfrak{S}_r$ the *symmetric group* of degree r (the set of all $r!$ permutations) acting on the set $\{1, 2, \dots, r\}$. For each permutation $\sigma \in \mathfrak{S}_r$, $\epsilon(\sigma)$ denotes its signature ± 1:

$$\epsilon(\sigma) = (-1)^{\ell(\sigma)}, \quad \ell(\sigma) = \#\{(i,j); i < j,\ \sigma(i) > \sigma(j)\}. \tag{6.4}$$

Here, $\ell(\sigma)$ stands for the *number of inversions* in σ, namely, the number of all pairs (i,j) $(i < j)$ whose ordering is reversed by σ.

It is sometimes more convenient to use sets of indices instead of sequences of indices. When we use the notation $\xi^I_J(X)$ with some sets I, J of indices, we define

$$\xi^I_J(X) = \xi^{i_1\dots i_r}_{j_1\dots j_r}(X), \tag{6.5}$$

assuming that $|I| = |J| = r$, and replacing the elements of I and J by the increasing sequences $i_1 < \ldots < i_r$ and $j_1 < \ldots < j_r$, respectively. Note that, for a given $m \times n$ matrix X, we have essentially $\binom{m}{r} \times \binom{n}{r}$ minor determinants $\xi^I_J(X) = \xi^{i_1 \ldots i_r}_{j_1 \ldots j_r}(X)$ of degree r. We will also write simply $\xi^{i_1 \ldots i_r}_{j_1 \ldots j_r}$ or ξ^I_J, when the matrix is clearly recognized from the context.

Among many others, we recall one important fact concerning matrices and determinants. Let X be an $n \times m$ matrix, and Y an $m \times n$ matrix, so that the product XY becomes an $n \times n$ square matrix. Then the determinant of XY is expressed as follows in terms of the minor determinants of X and Y: When $m < n$, the rank of the product

$$XY = \begin{bmatrix} x_{11} \ldots x_{1m} \\ \vdots \quad \vdots \\ \\ x_{n1} \ldots x_{nm} \end{bmatrix} \begin{bmatrix} y_{11} & \ldots & y_{1n} \\ \vdots & & \vdots \\ y_{m1} & \ldots & y_{mn} \end{bmatrix} \tag{6.6}$$

is less than n, and hence $\det(XY) = 0$. If $m > n$ and

$$XY = \begin{bmatrix} x_{11} & \ldots & x_{1m} \\ \vdots & & \vdots \\ x_{n1} & \ldots & x_{nm} \end{bmatrix} \begin{bmatrix} y_{11} \cdots y_{1n} \\ \\ \vdots \quad \vdots \\ \\ y_{m1} \cdots y_{mn} \end{bmatrix}, \tag{6.7}$$

then we have

$$\det(XY) = \sum_{k_1 < \cdots < k_n} \xi^{1 \ \ldots \ n}_{k_1 \ldots k_n}(X)\, \xi^{k_1 \ldots k_n}_{1 \ \ldots \ n}(Y). \tag{6.8}$$

(A proof of this can be provided by using the exterior algebra. ⇒ Appendix 10: Application of the exterior algebra.) This fact can be reformulated as follows into a theorem concerning general minor determinants.

THEOREM 6.1 (Minor determinants of the product of matrices). *Let X be an $l \times m$ matrix, and Y an $m \times n$ matrix. Then the minor determinants of the product XY of matrices are given as follows: For any choice of row indices $i_1, \ldots, i_r \in \{1, \ldots, l\}$ and column indices $j_1, \ldots, j_r \in \{1, \ldots, n\}$,*

$$\xi^{i_1 \ldots i_r}_{j_1 \ldots j_r}(XY) = \sum_{k_1 < \ldots < k_r} \xi^{i_1 \ldots i_r}_{k_1 \ldots k_r}(X)\, \xi^{k_1 \ldots k_r}_{j_1 \ldots j_r}(Y). \tag{6.9}$$

The summation is taken over all increasing sequences $k_1 < \ldots < k_r$ of length r in $\{1, \ldots, m\}$.

It is easy to remember this formula, since it can be understood as a generalization of the formula for the product of matrices, and contains no minus signs. With the notation of indexing sets, it becomes even simpler: For any index sets I, J with $|I| = |J| = r$, we have

$$\xi^I_J(XY) = \sum_{|K|=r} \xi^I_K(X)\, \xi^K_J(Y). \tag{6.10}$$

Relatively invariant minor determinants

The triangular matrices play a very important role in the theory of matrices and determinants, although this point may not be emphasized so much in ordinary textbooks on linear algebra.

Consider a lower triangular $m \times m$ matrix $A = (a_{ij})_{ij}$ ($a_{ij} = 0$ for $i < j$), and denote by $a_i = a_{ii}$ ($i = 1, \dots, m$) the diagonal entries of A. Multiplying an arbitrary $m \times n$ matrix X by A from the left, we examine the minor determinants of AX:

$$\xi^{1\dots r}_{j_1\dots j_r}(AX) = \sum_{k_1<\dots<k_r} \xi^{1\dots r}_{k_1\dots k_r}(A)\, \xi^{k_1\dots k_r}_{j_1\dots j_r}(X). \tag{6.11}$$

Since A is lower triangular, we have $\xi^{1\dots r}_{1\dots r}(A) = a_1 \dots a_r$; for the other sequences of indices $k_1 < \dots < k_r$, we have $\xi^{1\dots r}_{k_1\dots k_r}(A) = 0$:

$$A = \begin{array}{c} \begin{array}{cccc} k_1 & & k_2 & \cdots \end{array} \\ \begin{bmatrix} a_1 & & & \\ * & a_2 & & 0 \\ * & * & a_3 & \\ \vdots & \vdots & \vdots & \ddots \end{bmatrix} \end{array} \quad \Rightarrow \quad A^{1\ \cdots\ r}_{k_1\dots k_r} = \begin{array}{c} \begin{array}{cccc} k_1 & k_2 & \cdots & \end{array} \\ \begin{bmatrix} a_1 & & & \\ * & 0 & & 0 \\ * & a_3 & 0 & \\ \vdots & \vdots & \vdots & \ddots \end{bmatrix} \end{array} \tag{6.12}$$

Hence we have

$$\xi^{1\dots r}_{j_1\dots j_r}(AX) = a_1 \cdots a_r\, \xi^{1\dots r}_{j_1\dots j_r}(X). \tag{6.13}$$

We have the same phenomenon with respect to the multiplication by an upper triangular matrix from the right, as well.

PROPOSITION 6.2. *Let X be an $m \times n$ matrix.*
(1) *Let A be an $m \times m$ lower triangular matrix with diagonal entries $a_1, \dots, a_m$. Then for any $j_1, \dots, j_r$, we have*

$$\xi^{1\dots r}_{j_1\dots j_r}(AX) = a_1 \cdots a_r\, \xi^{1\dots r}_{j_1\dots j_r}(X). \tag{6.14}$$

(2) *Let B an $n \times n$ upper triangular matrix with diagonal entries $b_1, \dots, b_n$. Then for any $i_1, \dots, i_r$, we have*

$$\xi^{i_1\dots i_r}_{1\dots r}(XB) = b_1 \cdots b_r\, \xi^{i_1\dots i_r}_{1\dots r}(X). \tag{6.15}$$

Under the left multiplication by lower triangular matrices, the minor determinants $\xi^{1\dots r}_{j_1\dots j_r}$, with row indices $1, \dots, r$ (packed upwards), transform in a simple manner by picking up the diagonal entries; we refer to this property as the *relative invariance*. In particular, they are invariant with respect to left multiplication by lower triangular matrices whose diagonal entries are all 1. As for right multiplication by upper triangular matrices, the minor determinants $\xi^{i_1\dots i_r}_{1\dots r}$ with column indices $1, \dots, r$ (packed leftwards) carry the relative invariance.

6.2. Gauss decomposition and the Jacobi identity

The *Gauss decomposition* is a way of expressing a matrix as the product of a lower triangular matrix and an upper triangular matrix; it is also called simply the *triangular decomposition.* In the case of 2×2 matrices, it refers to the decomposition

$$\begin{bmatrix} a & b \\ c & d \end{bmatrix} = \begin{bmatrix} 1 & 0 \\ ca^{-1} & 1 \end{bmatrix} \begin{bmatrix} a & b \\ 0 & d - ca^{-1}b \end{bmatrix}. \tag{6.16}$$

There is nothing special about the Gauss decomposition, since it is a reformulation of the elimination (sweeping out) method for matrices. But it provides a strikingly powerful tool, as we will see below.

Suppose now that an $n \times n$ matrix $X = (x_{ij})_{ij}$ is given, and that the $(1,1)$-component x_{11} is nonzero. Using x_{11} as the pivot, we can change all the other entries in the first column into zero, by a repetition of elementary transformations with respect to rows:

$$X = \begin{bmatrix} x_{11} & x_{12} & \cdots & \\ x_{21} & x_{22} & \cdots & \\ \vdots & \vdots & & \\ x_{n1} & x_{2n} & & \end{bmatrix} \quad \longrightarrow \quad X' = \begin{bmatrix} x_{11} & x_{12} & \cdots & \\ 0 & x'_{22} & \cdots & \\ \vdots & \vdots & & \\ 0 & x'_{2n} & & \end{bmatrix}. \tag{6.17}$$

If $x'_{22} \neq 0$ at this step, we can use it again as the pivot to eliminate the entries below x'_{22} in the second column:

$$X' = \begin{bmatrix} x_{11} & x_{12} & \cdots & \\ 0 & x'_{22} & \cdots & \\ 0 & x'_{23} & \cdots & \\ \vdots & \vdots & & \end{bmatrix} \quad \longrightarrow \quad X'' = \begin{bmatrix} x_{11} & x_{12} & x_{13} & \cdots & \\ 0 & x'_{22} & x'_{23} & \cdots & \\ 0 & 0 & x''_{33} & \cdots & \\ \vdots & \vdots & & & \end{bmatrix}. \tag{6.18}$$

In repeating this procedure, suppose that the diagonal entry in the next column is nonzero at each step. Then we would finally arrive at an upper triangular matrix. Note that we are using only the procedure of adding a multiple of some row to another row downward. All the elementary transformations we have used can be represented together by a single lower triangular matrix P whose diagonal entries are all 1:

$$PX = \begin{bmatrix} x_{11} & x_{12} & x_{13} & \cdots \\ & x'_{22} & x'_{23} & \cdots \\ & & x''_{33} & \cdots \\ 0 & & & \end{bmatrix} \tag{6.19}$$

Denoting the right-hand side by Z, we can write $X = P^{-1}Z$, so that the matrix X is expressed as the product of a lower triangular matrix P^{-1} and an upper triangular matrix Z. We may expect that a *generic* matrix X can be decomposed in such a triangular form. In fact, if we choose X at random, we could probably expect that $x_{11} \neq 0$, and also at the next step, $x'_{12} \neq 0$, $\cdots$.

The procedure mentioned above can be formulated as the following theorem; we can also specify which matrices are decomposable in this way. In order to make statements simpler, let us introduce the phrase "unitriangular"; we say that a lower (or upper) triangular matrix is *unitriangular* if its diagonal entries are all 1.

THEOREM 6.3 (Gauss decomposition). *Let* $X = (x_{ij})_{i,j}$ *be an* $n \times n$ *square matrix satisfying the following conditions concerning the minor determinants*:

$$\xi_1^1 \neq 0, \quad \xi_{12}^{12} \neq 0, \quad \ldots, \xi_{1\ldots n}^{1\ldots n} \neq 0. \tag{6.20}$$

Then X *is expressed uniquely in the form*

$$X = WZ : \quad \begin{array}{l} (1)\ W \text{ is lower unitriangular,} \\ (2)\ Z \text{ is upper triangular.} \end{array} \tag{6.21}$$

The matrix Z can be decomposed further into the product of a diagonal matrix and an upper unitriangular matrix. Then X is decomposed into three parts:

$$X = X_- X_0 X_+ : \quad \begin{array}{l} (1)\ X_- \text{ is lower unitriangular,} \\ (2)\ X_0 \text{ is diagonal,} \\ (3)\ X_+ \text{ is upper unitriangular.} \end{array} \tag{6.22}$$

Imagining that X is like a fish lying on a cutting board, X is being sliced into three parts: the lower part X_- below the backbone, the middle backbone X_0, and the upper part X_+ above the backbone[1]. When we cut X into two slices as in the theorem, the backbone is attached to either the upper or the lower slice.

We now give the outline of the proof. It is not difficult to decompose X into the form

$$\begin{bmatrix} 1 & 0 & \ldots & 0 \\ x_{21}/x_{11} & & & \\ \vdots & & I & \\ x_{n1}/x_{11} & & & \end{bmatrix} \begin{bmatrix} 1 & 0 & \ldots & 0 \\ 0 & & & \\ \vdots & & Y & \\ 0 & & & \end{bmatrix} \begin{bmatrix} x_{11} & x_{12} & \ldots & x_{1n} \\ 0 & & & \\ \vdots & & I & \\ 0 & & & \end{bmatrix}. \tag{6.23}$$

(Here, I denotes the identity matrix of degree $n-1$, and Y is an $(n-1) \times (n-1)$ matrix.) If we express this decomposition as $X = AX'B$, from Proposition 6.2 we have

$$\xi_{1\ldots r}^{1\ldots r}(X) = \xi_{1\ldots r}^{1\ldots r}(X')x_{11} = \xi_{1\ldots r-1}^{1\ldots r-1}(Y)\, x_{11}. \tag{6.24}$$

Hence $\xi_{1\ldots r}^{1\ldots r}(Y) \neq 0$ for $r = 1, \ldots, n-1$. So we can proceed by induction. It is easy to prove the uniqueness of the decomposition. Suppose that we have two decompositions $X = W_1 Z_1 = W_2 Z_2$. Then $W_2^{-1} W_1 = Z_2 Z_1^{-1}$. The left-hand side of this expression is lower triangular with 1's on the diagonal, and the right-hand side is upper triangular. Hence we have $W_2^{-1} W_1 = 1$, $Z_2 Z_1^{-1} = 1$, namely $W_1 = W_2$ and $Z_1 = Z_2$.

Remark: For a general $n \times n$ matrix X, consider the expansion of the determinant $\det(X) = \xi_{1\ldots n}^{1\ldots n}$ along the n-th column:

$$\xi_{1\ldots n}^{1\ldots n} = \sum_{i=1}^{n} (-1)^{n-i} x_{in}\, \xi_{1\ldots n-1}^{1\ldots \hat{i} \ldots n}. \tag{6.25}$$

[1] According to Kanehisa Takasaki, "The Toda lattice is a sort of fish. If we slice it open into two parts, then each of them is the KP." (*Mathematical Physics and Professor Mikio Sato, II*, The Joy of Mathematics No. 14, pp 63–72, Nihon-Hyoronsha, 1999, in Japanese)

From this expression we see that if $\xi_{1\dots n}^{1\dots n} \neq 0$, then there is an index i such that $\xi_{1\dots n-1}^{1\dots\hat{i}\dots n} \neq 0$. Setting $i = \sigma(n)$, we repeat the same procedure for minor determinants. Then we can find a permutation $\sigma \in \mathfrak{S}_n$ such that

$$\xi_1^{\sigma(1)}, \quad \xi_{12}^{\sigma(1)\sigma(2)}, \quad \dots, \quad \xi_{1\dots n}^{\sigma(1)\dots\sigma(n)} \tag{6.26}$$

are all nonzero. If we denote by

$$E_\sigma = \sum_{j=1}^{n} E_{\sigma(j)j} \quad (E_{ij} \text{ are the matrix units}) \tag{6.27}$$

the permutation matrix corresponding to σ, then the product $E_\sigma X$ satisfies the condition of the theorem above. Hence, for each σ chosen as above, we obtain the decomposition $X = E_{\sigma^{-1}}WZ$.

Denoting a general $n \times n$ matrix by $X = (x_{ij})_{ij}$, we now regard the entries x_{ij} as the coordinate functions of the matrix space. Then the minor determinants $\xi_1^1, \dots, \xi_{1\dots n}^{1\dots n}$ are all nonzero functions. In this sense, X has the Gauss decomposition, $X = WZ$, and the entries of $W = (w_{ij})_{ij}$ and $Z = (z_{ij})_{ij}$ are determined uniquely as rational functions in the variables x_{ij}. Let us denote by $w_{ij}(X)$ and $z_{ij}(X)$ those w_{ij} and z_{ij} regarded as functions in the entries of X, respectively. Then the Gauss decomposition of a generic matrix is expressed in the form

$$X = W(X)\,Z(X); \quad W(X) = (w_{ij}(X))_{ij}, \quad Z(X) = (z_{ij}(X))_{ij}. \tag{6.28}$$

From the uniqueness of the Gauss decomposition, we realize that each rational function $z_{ij}(X)$ is invariant under left multiplication by lower unitriangular matrices A. In fact, the Gauss decomposition of the product AX is given by $AX = (AW)Z$. Hence we have $Z(AX) = Z = Z(X)$, namely,

$$z_{ij}(AX) = z_{ij}(X) \quad (1 \le i, j \le n). \tag{6.29}$$

Each $w_{ij}(X)$ also carries a similar invariance property.

Proposition 6.4. *Consider the $n \times n$ matrix $X = (x_{ij})_{ij}$, regarding the entries as indeterminates. Let $X = WZ$ be the Gauss decomposition of X in the sense of matrices of rational functions; the entries of $W = (w_{ij})_{ij}$ and $Z = (z_{ij})_{ij}$ are rational functions in x_{ij}.*
(1) Each $z_{ij} = z_{ij}(X)$ is invariant under left multiplication by lower unitriangular matrices A: $z_{ij}(AX) = z_{ij}(X)$.
(2) Each $w_{ij} = w_{ij}(X)$ is invariant under right multiplication by invertible upper triangular matrices B: $w_{ij}(XB) = w_{ij}(X)$.

If we examine the inductive proof of the existence of the Gauss decomposition, we can immediately read off the diagonal entires of the upper triangular matrix Z:

$$z_{11} = \xi_1^1, \quad z_{22} = \frac{\xi_{12}^{12}}{\xi_1^1}, \quad \dots, \quad z_{nn} = \frac{\xi_{1\,\dots\,n}^{1\,\dots\,n}}{\xi_{1\dots n-1}^{1\dots n-1}}. \tag{6.30}$$

These are certainly invariants in the sense of the proposition above. Note also that the product of all these functions recovers the determinant of the original Z: $\xi_{1\dots n}^{1\dots n} = \det(X)$. Then let us ask: How can one express the general entries of the two matrices of the Gauss decomposition as rational functions in x_{ij}? The answer is given by the following explicit formula. We will prove it by using the invariance property with respect to triangular matrices.

THEOREM 6.5 (Explicit formula for the Gauss decomposition). *Let $X = (x_{ij})_{ij}$ be an $n \times n$ square matrix satisfying the condition $\xi^{1\dots r}_{1\dots r} \neq 0$ $(r = 1, \dots, n)$. Then each entry of the matrices $W = (w_{ij})_{ij}$ and $Z = (z_{ij})_{ij}$ in the Gauss decomposition $X = WZ$ of Theorem 6.3 is expressed as follows as a ratio of minor determinants of X:*

$$w_{ij} = \frac{\xi^{1\dots j-1\, i}_{1\dots j-1\, j}}{\xi^{1\dots j}_{1\dots j}}, \quad z_{ij} = \frac{\xi^{1\dots i-1\, i}_{1\dots i-1\, j}}{\xi^{1\dots i-1}_{1\dots i-1}} \qquad (1 \leq i, j \leq n). \tag{6.31}$$

Proof. Fixing the indices i, j, we compare $w_{ij} = w_{ij}(X)$ with the rational function

$$\varphi(X) = \frac{\xi^{1\dots j-1\, i}_{1\dots j-1\, j}(X)}{\xi^{1\dots j}_{1\dots j}(X)} \tag{6.32}$$

in the x variables. For any upper triangular matrix B with diagonal entries $b_1, \dots, b_n$, we have

$$\xi^{k_1\dots k_j}_{1\dots j}(XB) = \xi^{k_1\dots k_j}_{1\dots j}(X) b_1 \dots b_j. \tag{6.33}$$

This implies that the numerator and the denominator of $\varphi(X)$ have the same relative invariance with respect to the right action of B, and hence $\varphi(X)$ is invariant under such transformations. We now use the Gauss decomposition $X = WZ$. Since Z is upper triangular, we have

$$\varphi(X) = \varphi(WZ) = \varphi(W). \tag{6.34}$$

On the other hand, for the lower triangular matrix W, we have $\xi^{1\dots j-1 k}_{1\dots j-1\, j}(W) = w_{kj}$. Hence,

$$\varphi(X) = \varphi(W) = w_{ij} = w_{ij}(X). \tag{6.35}$$

This means that w_{ij} and φ coincide with each other as rational functions. We can apply a similar argument to Z. In the decomposition $X = WZ$, we can use the invariance with respect to W to show

$$\frac{\xi^{1\dots i-1\, i}_{1\dots i-1\, j}(X)}{\xi^{1\dots i-1}_{1\dots i-1}(X)} = \frac{\xi^{1\dots i-1\, i}_{1\dots i-1\, j}(Z)}{\xi^{1\dots i-1}_{1\dots i-1}(Z)} = \frac{z_{11} \dots z_{i-1\, i-1} z_{ij}}{z_{11} \dots z_{i-1\, i-1}} = z_{ij} = z_{ij}(X). \tag{6.36}$$

Hence, we have $z_{ij} = \xi^{1\dots i-1\, i}_{1\dots i-1\, j} / \xi^{1\dots i-1}_{1\dots i-1}$. (Q.E.D.)

In the decomposition $X = X_- X_0 X_+$ into three slices, the matrices $X_\pm = (x^\pm_{ij})_{ij}$ and $X_0 = \mathrm{diag}(x^0_1, \dots, x^0_n)$ are determined as follows:

$$x^-_{ij} = \frac{\xi^{1\dots j-1\, i}_{1\dots j-1\, j}}{\xi^{1\dots j}_{1\dots j}}, \quad x^0_i = \frac{\xi^{1\dots i}_{1\dots i}}{\xi^{1\dots i-1}_{1\dots i-1}}, \quad x^+_{ij} = \frac{\xi^{1\dots i-1\, i}_{1\dots i-1\, j}}{\xi^{1\dots i}_{1\dots i}} \qquad (1 \leq i, j \leq n). \tag{6.37}$$

Sometimes we need the explicit formula for the entries of the inverse matrix of $W = X_-$. By a method similar to that in the proof of the theorem, we can prove

$$(W^{-1})_{ij} = (-1)^{i-j} \frac{\xi^{1\dots \widehat{j} \dots i}_{1\ \dots\ i-1}}{\xi^{1\dots i-1}_{1\dots i-1}} \qquad (i \geq j), \tag{6.38}$$

where $1 \dots \widehat{j} \dots i$ is an abbreviation for the sequence $1 \dots j-1, j+1 \dots i$ obtained by skipping j. Naturally, this formula gives 0 for $i < j$, and 1 for $i = j$.

As for the invariants with respect to the action of triangular matrices, we have the following general theorem.

THEOREM 6.6 (Invariants under the action of triangular matrices). *Let $\varphi(X)$ be a rational function in x_{ij} with poles only along $\xi_1^1 \xi_{12}^{12} \cdots \xi_{1\dots n}^{1\dots n} = 0$. Then the following two conditions are equivalent*:

(*a*) *$\varphi(X)$ is invariant with respect to left multiplication by any lower unitriangular matrix A: $\varphi(AX) = \varphi(X)$.*
(*b*) *$\varphi(X)$ is a polynomial in $z_{ii}^{\pm 1} = z_{ii}(X)^{\pm 1}$ and $z_{ij} = z_{ij}(X)$ $(i < j)$.*

This implication $(b) \Rightarrow (a)$ is obvious from the invariance of $z_{ij}(X)$. As for the other implication $(a) \Rightarrow (b)$, by the decomposition $X = WZ$ we have

$$\varphi(X) = \varphi(WZ) = \varphi(Z) = \varphi(Z(X)). \tag{6.39}$$

The restriction of $\varphi(X)$ to the space of invertible upper triangular matrices becomes a polynomial in the variables $z_{ii}^{\pm 1}$ and z_{ij} $(i < j)$.

We now apply the idea of Gauss decomposition to proving the *Jacobi identity* concerning the inductive structure of the minor determinants. (The Gauss decomposition and the Jacobi identity are two sides of a coin, so to speak.) Algebraic relations among minor determinants, such as the Jacobi identity, play important roles in various aspects of mathematics, and also have many applications.

THEOREM 6.7 (Jacobi identity). *For minor determinants of an $n \times n$ square matrix X, we have*

$$\xi_{1\dots n-2}^{1\dots n-2} \xi_{1\dots n}^{1\dots n} = \xi_{1\dots n-1}^{1\dots n-1} \xi_{1\dots n-2,n}^{1\dots n-2,n} - \xi_{1\dots n-2,n}^{1\dots n-2,n-1} \xi_{1\dots n-2,n-1}^{1\dots n-2,n}. \tag{6.40}$$

Note that the left-hand side is the product of two minor determinants of degree $(n-2)$ and of degree n, while the right-hand side is a quadratic formula of minor determinants of degree $(n-1)$. Also, if we look only at the last two indices, it has the same structure as the determinant of 2×2 matrices. By permuting the indices, the Jacobi identity can be rewritten into various forms. The one concerning the first and the last indices $1, n$ (in the form of a window-frame) looks nice too:

$$\xi_{2\dots n-1}^{2\dots n-1} \xi_{1\dots n}^{1\dots n} = \xi_{1\dots n-1}^{1\dots n-1} \xi_{2\dots n}^{2\dots n} - \xi_{2\dots\, n}^{1\dots n-1} \xi_{1\dots n-1}^{2\dots\, n}. \tag{6.41}$$

This form of the Jacobi identity is often called the *Lewis Carroll formula.* (Lewis Carroll is the author of *Alice in Wonderland.*)

As we will see below, this formula of Jacobi (or Lewis Carroll) is already embedded in the explicit formula for the Gauss decomposition. By using the decomposition $X = WZ$, we can express the minor determinants of X in terms of the minor determinants of W and Z:

$$\xi_{j_1 \dots j_r}^{i_1 \dots i_r}(X) = \sum_{k_1 < \dots < k_r} \xi_{k_1 \dots k_r}^{i_1 \dots i_r}(W)\, \xi_{j_1 \dots j_1}^{k_1 \dots k_r}(Z). \tag{6.42}$$

Recall that, on the right-hand side, the entries of W and Z are expressed by means of the minor determinants of X. In this way, the formulas above can be regarded as providing various algebraic relations among the minor determinants of X. Let us look at the particular case $r = n-1$:

$$\xi^{1\dots n-2,n}_{1\dots n-2,n}(X) = \sum_{k_1<\dots<k_r} \xi^{1\dots n-2,n}_{k_1\dots k_r}(W)\,\xi^{k_1\dots k_r}_{1\dots n-2,n}(Z). \tag{6.43}$$

On the right-hand side, nontrivial terms arise from the following sets of indices:

(6.44)

$(k_1,\dots,k_n)$	W side	Z side
$(1,\dots,n-2,n-1)$	$w_{n,n-1}$	$z_1\cdots z_{n-2}z_{n-1,n}$
$(1,\dots,n-2,n)$	1	$z_1\cdots z_{n-2}z_n$

Here we have set $z_i = z_{ii}$ for simplicity. Rewriting the entries of W and Z by means of minor determinants of X, we obtain

$$\begin{aligned}\xi^{1\dots n-2,n}_{1\dots n-2,n} &= w_{n,n-1}z_1\cdots z_{n-2}z_{n-1,n} + z_1\cdots z_{n-2}z_n \\ &= \frac{\xi^{1\dots n-2,n}_{1\dots n-2,n-1}}{\xi^{1\dots n-1}_{1\dots n-1}}\,\xi^{1\dots n-2,n-1}_{1\dots n-2,n} + \xi^{1\dots n-2}_{1\dots n-2}\,\frac{\xi^{1\dots n}_{1\dots n}}{\xi^{1\dots n-1}_{1\dots n-1}},\end{aligned} \tag{6.45}$$

which is equivalent to

$$\xi^{1\dots n-1}_{1\dots n-1}\xi^{1\dots n-2,n}_{1\dots n-2,n} - \xi^{1\dots n-2,n}_{1\dots n-2,n-1}\xi^{1\dots n-2,n-1}_{1\dots n-2,n} = \xi^{1\dots n-2}_{1\dots n-2}\xi^{1\dots n}_{1\dots n}. \tag{6.46}$$

This is the Jacobi identity we expected.

6.3. Diagonalization of triangular matrices

We discuss another explicit formula which will be used in later arguments. Let $M = (f_{ij})_{ij}$ be an $n\times n$ upper triangular matrix, and write the diagonal entries as $\varepsilon_i = f_{ii}$ $(i = 1,\dots,n)$.

$$M = \begin{bmatrix} \varepsilon_1 & f_{12} & f_{13} & \cdots & f_{1n} \\ & \varepsilon_2 & f_{23} & \cdots & f_{2n} \\ & & \ddots & & \vdots \\ & 0 & & \varepsilon_{n-1} & f_{n-1\,n} \\ & & & & \varepsilon_n \end{bmatrix}. \tag{6.47}$$

If all the eigenvalues $\varepsilon_1,\dots,\varepsilon_n$ are mutually distinct, such a matrix can be diagonalized by using an upper unitriangular matrix $P = (p_{ij})_{ij}$:

$$M = P\,D\,P^{-1}, \quad \text{namely} \quad MP = PD. \tag{6.48}$$

Note also that $D = \mathrm{diag}(\varepsilon_1,\dots,\varepsilon_n)$, and that the matrix P is determined uniquely from M. In this section we discuss how the entries of P are determined from those of M.

We remark first that, from the triangularity of M and P, the j-th column of P is determined by the system of linear equations

$$\begin{bmatrix} \varepsilon_1 & f_{12} & \cdots & f_{1j} \\ & \ddots & & \vdots \\ & & \varepsilon_{j-1} & f_{j-1\,j} \\ 0 & & & \varepsilon_j \end{bmatrix} \begin{bmatrix} p_{1j} \\ \vdots \\ p_{j-1\,j} \\ 1 \end{bmatrix} = \begin{bmatrix} p_{1j} \\ \vdots \\ p_{j-1\,j} \\ 1 \end{bmatrix} \varepsilon_j. \tag{6.49}$$

By decomposing the matrices into blocks according to the sum $j = (j-1)+1$, we see that the system above is equivalent to

$$\begin{bmatrix} \varepsilon_1-\varepsilon_j & f_{12} & \cdots & f_{1\,j-1} \\ & \ddots & & \vdots \\ & & \varepsilon_{j-2}-\varepsilon_j & f_{j-2\,j-1} \\ 0 & & & \varepsilon_{j-1}-\varepsilon_j \end{bmatrix} \begin{bmatrix} p_{1j} \\ \vdots \\ p_{j-2\,j} \\ p_{j-1\,j} \end{bmatrix} = - \begin{bmatrix} f_{1j} \\ \vdots \\ f_{j-2\,j} \\ f_{j-1\,j} \end{bmatrix}. \tag{6.50}$$

Since $\prod_{1\le k<j}(\varepsilon_k - \varepsilon_j) \neq 0$, the matrix on the left-hand side has nonzero determinant, and hence the p_{ij} are determined by the Cramer formula. In the computation of the Cramer formula, we insert the column vector on the right-hand side into the i-th column of the matrix on the left-hand side. If we transfer the column vector further to the $(j-1)$-th column, as a result we obtain the following expression for p_{ij}.

THEOREM 6.8. *Suppose that the diagonal entries $\varepsilon_1, \ldots, \varepsilon_n$ are mutually distinct. Then the matrix M in (6.47) is diagonalized by a unique upper unitriangular matrix $P = (p_{ij})_{ij}$:*

$$M = PDP^{-1}, \qquad D = \operatorname{diag}(\varepsilon_1, \ldots, \varepsilon_n). \tag{6.51}$$

Furthermore, the entries p_{ij} $(i \le j)$ of P are expressed as follows in terms of determinants:

$$p_{ij} = \frac{(-1)^{j-i}}{\displaystyle\prod_{i\le k<j}(\varepsilon_k - \varepsilon_j)} \det \begin{bmatrix} f_{i\,i+1} & f_{i\,i+2} & \cdots & f_{i\,j} \\ \varepsilon_{i+1\,j} & f_{i+1\,i+2} & \cdots & f_{i+1\,j} \\ & \ddots & \ddots & \vdots \\ 0 & & \varepsilon_{j-1\,j} & f_{j-1\,j} \end{bmatrix}, \tag{6.52}$$

where $\varepsilon_{ij} = \varepsilon_i - \varepsilon_j$.

Along the lower subdiagonal, we arrange ε_{kj} $(k = i+1, i+2, \ldots, j-1)$ in this order.

This determinant formula can be compared with that of (5.65) in the setting of the Schur functions $S_\lambda(\boldsymbol{t})$ of Section 5.4. We also have an expansion formula for p_{ij} $(i<j)$, corresponding to the formula (5.57) for $p_l(\boldsymbol{t})$:

$$p_{ij} = \sum_{r=1}^{j-i} (-1)^r \sum_{i=i_0<i_1<\cdots<i_r=j} \frac{f_{i_0 i_1}}{\varepsilon_{i_0 j}} \cdots \frac{f_{i_{r-1} i_r}}{\varepsilon_{i_{r-1} j}}. \tag{6.53}$$

(Hint for a proof: Fixing i, j with $i<j$, consider the upper unitriangular matrix whose (a,b)-components are f_{ab}/ε_{aj} $(i \le a < b \le j)$. Then compute the (i,j)-component of its inverse matrix by using the Neumann series.) Conversely, f_{ij} $(i<j)$ are expressed in terms of the p_{ij} as follows:

$$f_{ij} = \sum_{r=1}^{j-i} (-1)^r \sum_{i=i_0<i_1<\cdots<i_r=j} \varepsilon_{i,i_1} p_{i_0 i_1} p_{i_1 i_2} \cdots p_{i_{r-1} i_r}. \tag{6.54}$$

6.4. Plücker relations

In Theorem 6.1, we presented a formula for the minor determinants of the product of matrices. Another fundamental formula, which we are going to discuss now, is the one concerning the division of row indices (or column indices) into two parts. A typical situation is the following: Given an $n \times n$ matrix $X = (x_{ij})_{ij}$, divide the n rows of X into the two parts, the first r rows and the other s rows $(r + s = n)$. Setting

$$\begin{bmatrix} x_{11} & \cdots & x_{1n} \\ \vdots & & \vdots \\ x_{n1} & \cdots & x_{nn} \end{bmatrix} = \begin{bmatrix} y_{11} & \cdots & y_{1n} \\ \vdots & & \vdots \\ y_{r1} & \cdots & y_{rn} \\ z_{11} & \cdots & z_{1n} \\ \vdots & & \vdots \\ z_{s1} & \cdots & z_{sn} \end{bmatrix}, \tag{6.55}$$

we denote by Y the upper $r \times n$ matrix, and by Z the lower $s \times n$ matrix. Then the determinant of X is expressed in the form

$$\det(X) = \sum_{I \cup J = \{1,\dots,n\}} \epsilon(I;J)\, \xi_I^{\{1\dots r\}}(Y)\, \xi_J^{\{1\dots s\}}(Z) \tag{6.56}$$

by using the minor determinants of Y and Z. The summation on the right-hand side is taken over all possible ways of dividing the indexing set into the disjoint union $I \cup J = \{1, \dots, n\}$ of two subsets such that $|I| = r$ and $|J| = s$. The symbol $\epsilon(I;J)$ denotes a sign ± 1 determined by the arrangement of the two subsets I, J. For any pair of disjoint subsets $I, J \subset \{1, \dots, n\}$, $\epsilon(I;J)$ is defined by

$$\epsilon(I;J) = (-1)^{\ell(I;J)}; \quad \ell(I;J) = \#\{(i,j) \in I \times J; i > j\}. \tag{6.57}$$

The number $\ell(I;J)$ is the *inversion number* with respect to I and J; it counts the number of pairs $(i,j) \in I \times J$ that are against our expectation that elements of I should be smaller than those of J. (For a proof of (6.56) $\Rightarrow$ Appendix 10: Application of the exterior algebra.) We now introduce a notation. Suppose that three sets of indices I, J, K ($|I| = r$, $|J| = s$, $|K| = r + s$) are given. Then by arranging the elements of I, J, K so that $i_1 < \cdots < i_r$, $j_1 < \cdots < j_s$, $k_1 < \cdots < k_{r+s}$, respectively, we define

$$\xi_K^{I,J} = \xi_{k_1 \cdots k_{r+s}}^{i_1 \cdots i_r j_1 \cdots j_s}. \tag{6.58}$$

This definition is equivalent to

$$\xi_K^{I,J} = \begin{cases} \epsilon(I;J)\, \xi_K^{I \cup J} & (I \cap J = \emptyset), \\ 0 & (I \cap J \neq \emptyset). \end{cases} \tag{6.59}$$

THEOREM 6.9. *Let X be an $m \times n$ matrix. Take three sets of indices I_1, I_2 and J such that*

$$\begin{aligned} &I_1, I_2 \subset \{1, \dots, m\}; \quad |I_1| = r_1, \quad |I_2| = r_2, \\ &J \subset \{1, \dots, n\}; \qquad |J| = r_1 + r_2. \end{aligned} \tag{6.60}$$

Then we have the following identity for the minor determinants of X:

$$\xi_J^{I_1, I_2} = \sum_{J_1 \cup J_2 = J} \epsilon(J_1; J_2)\, \xi_{J_1}^{I_1} \xi_{J_2}^{I_2}, \tag{6.61}$$

where the summation is taken over all divisions $J = J_1 \cup J_2$ of J such that $|J_1| = r_1$, $|J_2| = r_2$.

Although the content of this theorem is the same as that of (6.56), this formulation is more convenient for our purpose.

The Plücker relations, as well as the Jacobi identity, are amongst the most important algebraic relations of minor determinants. The first nontrivial Plücker relation is the one concerning the minor determinants of degree 2 of a general 2×4 matrix:

$$\xi_{12}^{12}\xi_{34}^{12} - \xi_{13}^{12}\xi_{24}^{12} + \xi_{14}^{12}\xi_{23}^{12} = 0. \tag{6.62}$$

Relations like

$$\xi_1^1\xi_{23}^{12} - \xi_2^1\xi_{13}^{12} + \xi_3^1\xi_{12}^{12} = 0, \tag{6.63}$$

which contain minor determinants of different degrees, are also regarded as a variant of Plücker relations. We formulate this class of quadratic relations in the most general form.

THEOREM 6.10 (Generalized Plücker relations). *Setting $r_1 + s_1 = l_1$, $r_2 + s_2 = l_2$, we take the following sets of indices:*

$$\begin{aligned} &I_1, I_2 \subset \{1, \dots, m\}; \qquad |I_1| = l_1, \quad |I_2| = l_2, \\ &J_1, J_2, K \subset \{1, \dots, n\}; \quad |J_1| = r_1, \quad |J_2| = r_2, \quad |K| = s_1 + s_2. \end{aligned} \tag{6.64}$$

If $|I_1 \cup I_2| < s_1 + s_2$, then the following quadratic relations hold for the minor determinants of a general $m \times n$ matrix X:

$$\sum_{K_1 \cup K_2 = K} \epsilon(K_1; K_2)\, \xi_{J_1, K_1}^{I_1} \xi_{K_2, J_2}^{I_2} = 0. \tag{6.65}$$

Consider the special case where $l_1 \le l_2$ and

$$I_1 = \{1, \dots, l_1\}, \quad I_2 = \{1, \dots, l_2\}, \quad |J_1| = l_1 - 1, \quad J_2 = \emptyset. \tag{6.66}$$

If we take the set K of indices to be shuffled as large as $|K| = l_2 + 1$, the assumption of the theorem is fulfilled. This particular case of the theorem can be restated as follows. Take two subsets $J, K \subset \{1, \dots, n\}$ such that $|J| = r - 1$, $|K| = l + 1$, assuming that $1 \le r \le l \le m$. Then we obtain the quadratic relation

$$\sum_{k \in K} \epsilon(\{k\}; K)\, \xi_{J,\ \{k\}}^{\{1 \dots r\}} \xi_{K \setminus \{k\}}^{\{1 \dots l\}} = 0. \tag{6.67}$$

In terms of the indices,

$$\sum_{\nu=0}^{l}(-1)^{\nu}\xi^{1\ \cdots\ r}_{j_1\cdots j_{r-1}\,k_\nu}\,\xi^{1\ \cdots\ l}_{k_0\ldots\widehat{k_\nu}\ldots k_l}=0. \tag{6.68}$$

The last formula is the *Plücker relation*, as it is usually called. It is also known that, if we regard the entries of $X=(x_{ij})_{ij}$ as variables, the algebraic relations among the minor determinants $\xi^{1\ldots r}_{j_1\ldots j_r}$ $(j_1<\cdots<j_r)$ are essentially exhausted by the Plücker relations described above.

Theorem 6.10 can be proved by using Theorem 6.9. In each term of the left-hand side of (6.65), we apply Theorem 6.9 to divide the two minor determinants with respect to rows:

$$\begin{aligned}
&\sum_{K_1\cup K_2=K}\epsilon(K_1;K_2)\,\xi^{I_1}_{J_1,K_1}\,\xi^{I_2}_{K_2,J_2}\\
&=\sum_{K_1\cup K_2=K}\epsilon(K_1;K_2)\sum_{\substack{L_1\cup M_1=I_1\\ M_2\cup L_2=I_2}}\epsilon(L_1;M_1)\epsilon(M_2;L_2)\xi^{L_1}_{J_1}\,\xi^{M_1}_{K_1}\xi^{M_2}_{K_2}\,\xi^{L_2}_{J_2}\\
&=\sum_{\substack{L_1\cup M_1=I_1\\ M_2\cup L_2=I_2}}\epsilon(L_1;M_1)\epsilon(M_2;L_2)\,\xi^{L_1}_{J_1}\,\xi^{L_2}_{J_2}\sum_{K_1\cup K_2=K}\epsilon(K_1;K_2)\,\xi^{M_1}_{K_1}\,\xi^{M_2}_{K_2}.
\end{aligned} \tag{6.69}$$

Since $|M_1|=s_1$, $|M_2|=s_2$, under the assumption $M_1\subset I_1$, $M_2\subset I_2$, $|I_1\cup I_2|<s_1+s_2$, we have $M_1\cap M_2\neq\emptyset$. Hence we obtain

$$\sum_{K_1\cup K_2=K}\epsilon(K_1;K_2)\,\xi^{M_1}_{K_1}\,\xi^{M_2}_{K_2}=\xi^{M_1,M_2}_{K}=0. \tag{6.70}$$

This gives (6.65), as desired. (Q.E.D.)

CHAPTER 7

Gauss Decomposition and Birational Transformations

In this chapter we explain in the framework of matrices a method for constructing birational transformations like those that appeared in the Bäcklund transformations for the Painlevé equations. We first deal with the case of $n \times n$ matrices. On the basis of this A_{n-1} case, we proceed to birational transformations of type A_∞, and of type $A^{(1)}_{n-1}$. The Jacobi-Trudi formula for the ϕ-factors then follows naturally from this construction.

7.1. Birational transformation of the f variables

We start with the following $n \times n$ upper triangular matrix M:

$$M = \begin{bmatrix} \varepsilon_1 & f_{12} & f_{13} & \cdots & f_{1n} \\ & \varepsilon_2 & f_{23} & \cdots & f_{2n} \\ & & \ddots & & \vdots \\ & 0 & & \varepsilon_{n-1} & f_{n-1\,n} \\ & & & & \varepsilon_n \end{bmatrix} = \sum_{j=1}^{n} \varepsilon_j E_{jj} + \sum_{i<j} f_{ij} E_{ij}. \tag{7.1}$$

We distinguish the diagonal entries from the others, denoting them as ε_j ($j = 1, \ldots, n$). We always assume that M has n mutually distinct eigenvalues, namely, $\varepsilon_i \neq \varepsilon_j$ ($i \neq j$). In the above, the E_{ij} stand for the matrix units; the (i,j)-component of E_{ij} is 1, and the other components are 0. In what follows, we refer to the space of all such matrices M tentatively as the M *space*, and regard the variables f_{ij} and ε_j as the coordinate functions of the M space. We set

$$\alpha_j = \varepsilon_j - \varepsilon_{j+1} \quad (j = 1, \ldots, n-1) \tag{7.2}$$

and

$$f_j = f_{j,j+1} \qquad (j = 1, \ldots, n-1). \tag{7.3}$$

These subdiagonal coordinates play a special role below.

We now explain a method for constructing birational transformations on the M space. For each index $k = 1, \dots, n-1$, we look for a matrix

$$
G_k = \begin{array}{c} \begin{array}{cc} \quad\; k & k+1 \end{array} \\ \begin{bmatrix} \ddots & & & \\ & 1 & & \\ & u_k & 1 & \\ & & & \ddots \end{bmatrix} \end{array} = 1 + u_k E_{k+1,k} \tag{7.4}
$$

with one nontrivial entry on the lower subdiagonal such that the *sandwich* $\widetilde{M} = G_k M G_k^{-1}$ again becomes upper triangular. Let us compute the block corresponding to the indices $k, k+1$:

$$
\begin{aligned}
&\begin{bmatrix} 1 & 0 \\ u_k & 1 \end{bmatrix} \begin{bmatrix} \varepsilon_k & f_k \\ 0 & \varepsilon_{k+1} \end{bmatrix} \begin{bmatrix} 1 & 0 \\ -u_k & 1 \end{bmatrix} \\
&\qquad = \begin{bmatrix} \varepsilon_k - u_k f_k & f_k \\ u_k(\varepsilon_k - \varepsilon_{k+1}) - u_k^2 f_k & \varepsilon_{k+1} + u_k f_k \end{bmatrix}.
\end{aligned} \tag{7.5}
$$

Hence, we have a nontrivial solution

$$
u_k = \frac{\varepsilon_k - \varepsilon_{k+1}}{f_k} = \frac{\alpha_k}{f_k} \tag{7.6}
$$

(provided that $f_k \neq 0$). With u_k thus defined, the right-hand side of (7.5) gives

$$
\begin{bmatrix} \varepsilon_{k+1} & f_k \\ 0 & \varepsilon_k \end{bmatrix}. \tag{7.7}
$$

In view of this computation, we set

$$
G_k = 1 + \frac{\alpha_k}{f_k} E_{k+1,k}. \tag{7.8}
$$

Then the conjugation $\widetilde{M} = G_k M G_k^{-1}$ of M by G_k is again an upper triangular matrix. We now denote the diagonal entries of $\widetilde{M} = G_k M G_k^{-1}$ by $\widetilde{\varepsilon}_j$, and the upper triangular entries by $\widetilde{f}_{ij}$ $(i < j)$. These new entries are computed as

$$
\widetilde{\varepsilon}_j = \varepsilon_j \ \ (j \neq k, k+1), \quad \widetilde{\varepsilon}_k = \varepsilon_{k+1}, \ \ \widetilde{\varepsilon}_{k+1} = \varepsilon_k \ \ (j \neq k, k+1) \tag{7.9}
$$

and

$$
\begin{aligned}
\widetilde{f}_{k,k+1} &= f_{k,k+1}, & & \\
\widetilde{f}_{k+1,j} &= f_{k+1,j} + \frac{\alpha_k}{f_k} f_{k,j} & & (j > k+1), \\
\widetilde{f}_{i,k} &= f_{i,k} - \frac{\alpha_k}{f_k} f_{i,k+1} & & (i < k), \\
\widetilde{f}_{ij} &= f_{ij} & & (\text{otherwise}).
\end{aligned} \tag{7.10}
$$

In particular, the subdiagonal entries $f_j = f_{j,j+1}$ are transformed by the formulas

$$
\widetilde{f}_{k-1} = f_{k-1} - \frac{\alpha_k}{f_k} f_{k-1,k+1}, \quad \widetilde{f}_k = f_k, \quad \widetilde{f}_{k+1} = f_{k+1} + \frac{\alpha_k}{f_k} f_{k,k+2}, \tag{7.11}
$$

while the other f_j $(j \neq k-1, k+1)$ remain invariant: $\widetilde{f}_j = f_j$. If $f_{ij} = 1$ for $j - i = 2$, then this is precisely the pattern of the rational transformation s_k that we have seen repeatedly in the Bäcklund transformations of the Painlevé equations!

For each $k = 1, 2, \ldots, n-1$, we define the s_k transformation of the variables ε_j, f_{ij} by

(7.12) $$s_k(\varepsilon_j) = \text{the } (j,j)\text{-component of the matrix } G_k M G_k^{-1} \quad (j = 1, \ldots, n),$$
$$s_k(f_{ij}) = \text{the } (i,j)\text{-component of the matrix } G_k M G_k^{-1} \quad (1 \le i < j \le n).$$

As before, given a function φ in the variables ε_j, f_{ij}, we use the symbol $s_k(\varphi)$ to refer to the function obtained from φ by replacing all the appearances of ε_j, f_{ij} by $s_k(\varepsilon_j)$, $s_k(f_{ij})$, respectively. Also, for a matrix $\Phi = (\varphi_{ij})_{ij}$ of such functions, we denote by $s_k(\Phi) = (s_k(\varphi_{ij}))_{ij}$ the matrix obtained by applying s_k to all the entries of Φ individually. With this abbreviation, s_k can be defined simply by the formula

(7.13) $$s_k(M) = G_k\, M\, G_k^{-1} \quad (k = 1, \ldots, n-1).$$

We remark that the action of s_k on the items (7.10) can be expressed all together by the formula

(7.14) $$s_k(f_{ij}) = f_{ij} + \frac{\alpha_k}{f_k}\,(\delta_{k+1,i} f_{k,j} - \delta_{j,k} f_{i,k+1}) \quad (i < j),$$

where δ_{ij} is the Kronecker delta, representing 1 for $i = j$, and 0 for $i \ne j$,

It turns out after all that the group $W = \langle\, s_1, \ldots, s_{n-1}\,\rangle$ generated by the rational transformations $s_1, \ldots, s_{n-1}$ defined above is a finite group isomorphic to the symmetric group $\mathfrak{S}_n$ of degree n, i.e. the Weyl group of type A_{n-1}. The Cartan matrix $A = (a_{ij})_{i,j=1}^{n-1}$ of this case is the symmetric matrix with (-1)'s on the subdiagonal beside the 2's on the diagonal; its Dynkin diagram is the chain of $n-1$ nodes $\circ$:

(7.15) $$A = \begin{bmatrix} 2 & -1 & 0 & \cdots & 0 \\ -1 & 2 & -1 & & 0 \\ 0 & -1 & 2 & & \\ \vdots & & & \ddots & \vdots \\ 0 & 0 & & \cdots & \end{bmatrix}$$

The corresponding Weyl group $W = \langle\, s_1, s_2, \ldots, s_{n-1}\,\rangle$ is defined by the fundamental relations

(7.16) $$s_i^2 = 1, \quad s_i s_j = s_j s_i \quad (|i-j| \ge 2), \quad s_i s_j s_i = s_j s_i s_j \quad (|i-j| = 1).$$

In the symmetric group $\mathfrak{S}_n$ of degree n acting on the index set $\{1, \ldots, n\}$, consider the adjacent transpositions of indices

(7.17) $$\sigma_i = (i\ i+1) \qquad (i = 1, \ldots, n-1).$$

Then it is well known that the correspondence $s_i \mapsto \sigma_i$ induces an isomorphism from W to $\mathfrak{S}_n$.

In our setting, the action of s_k on the coordinates ε_j $(j = 1, \ldots, n)$ of the diagonal part of M is given by

(7.18) $$s_k(\varepsilon_j) = \varepsilon_{\sigma_k(j)} \qquad (j = 1, \ldots, n);$$

this corresponds to the permutation of indices by the symmetric group (*permutation representation*). The action of s_k on $\alpha_j = \varepsilon_j - \varepsilon_{j+1}$ is the canonical action expressed as

(7.19) $$s_k(\alpha_j) = \alpha_j - \alpha_k a_{kj} \qquad (j = 1, \ldots, n-1)$$

in terms of the Cartan matrix.

THEOREM 7.1. *Using the matrix*

$$G_k = 1 + \frac{\alpha_k}{f_k} E_{k+1,k}, \tag{7.20}$$

define the birational transformation s_k $(k = 1, \ldots, n-1)$ *on the space of matrices* M (7.1) *through the conjugation*

$$s_k(M) = G_k \, M \, G_k^{-1} \tag{7.21}$$

of M *by* G_k. *Then the action of* s_k *on the coordinate functions* ε_j, f_{ij} $(i < j)$ *is given by*

$$\begin{aligned} s_k(\varepsilon_j) &= \varepsilon_{\sigma_k(j)} \qquad (j = 1, \ldots, n), \\ s_k(f_{ij}) &= f_{ij} + \frac{\alpha_k}{f_k} \left(\delta_{k+1,i} f_{k,j} - \delta_{j,k} f_{i,k+1} \right) \quad (1 \le i < j \le n). \end{aligned} \tag{7.22}$$

Furthermore, the group $W = \langle\, s_1, \ldots, s_{n-1} \,\rangle$ *generated by these birational transformations gives a realization of the Weyl group of type* A_{n-1} *(the symmetric group of degree n). Namely, the transformations* s_k *defined as above satisfy the commutation relations*

$$s_i^2 = 1, \quad s_i s_j = s_j s_i \ \ (|i-j| \ge 2), \quad s_i s_j s_i = s_j s_i s_j \ \ (|i-j| = 1). \tag{7.23}$$

At present, we will not give a proof of the statement that $s_1, \ldots, s_n$ satisfy the commutation relations (7.23). This will be made clear gradually in this chapter as our argument proceeds.

We remark that the action of s_k on f_{ij} is again related to the Poisson structure. With respect to the commutator of matrices, defined by $[X, Y] = XY - YX$, it is easily checked that the matrix units E_{ij} satisfy the commutation relation

$$[E_{ij}, E_{kl}] = \delta_{jk} E_{il} - \delta_{li} E_{kj}. \tag{7.24}$$

Also, the commutator satisfies the following *universal* commutation relation:

$$[X, [Y, Z]] + [Y, [Z, X]] + [Z, [X, Y]] = 0, \tag{7.25}$$

called the Jacobi rule. Hence, if we define a bracket $\{\,,\,\}$ among the variables f_{ij} $(i < j)$ by the same formula

$$\{f_{ij}, f_{kl}\} = \delta_{jk} f_{il} - \delta_{li} f_{kj} \tag{7.26}$$

as in (7.24), it satisfies the Jacobi rule for any linear functions in the variables f_{ij}. For general functions in f_{ij}, we define the bracket as

$$\begin{aligned} \{\varphi, \psi\} &= \sum_{i<j;k<l} \frac{\partial \varphi}{\partial f_{ij}} \{f_{ij}, f_{kl}\} \frac{\partial \psi}{\partial f_{kl}} \\ &= \sum_{i<j<k} f_{ik} \left(\frac{\partial \varphi}{\partial f_{ij}} \frac{\partial \psi}{\partial f_{jk}} - \frac{\partial \varphi}{\partial f_{jk}} \frac{\partial \psi}{\partial f_{ij}} \right) \end{aligned} \tag{7.27}$$

by using the derivation rule. Then this bracket satisfies the conditions of the Poisson bracket. With this Poisson bracket, the action of s_k on f_{ij} $(i < j)$ is expressed as

$$s_k(f_{ij}) = f_{ij} + \frac{\alpha_k}{f_k} \{f_k, f_{ij}\}. \tag{7.28}$$

7.2. Birational transformations arising from the Gauss decomposition

We next consider the following invertible upper triangular matrix Z:

$$Z = \begin{bmatrix} z_1 & z_{12} & z_{13} & \cdots & z_{1n} \\ & z_2 & z_{23} & \cdots & z_{2n} \\ & & \ddots & & \vdots \\ & 0 & & z_{n-1} & z_{n-1\,n} \\ & & & & z_n \end{bmatrix}. \tag{7.29}$$

We have denoted the diagonal entries as $z_j = z_{jj}$ $(j = 1, \ldots, n)$, assuming that $z_j \neq 0$ for all j. We refer to the space of invertible upper triangular matrices Z as the Z *space*, and regard z_j $(1 \leq j \leq n)$, z_{ij} $(1 \leq i < j \leq n)$ as the coordinate functions of the Z space. In this section, we construct birational transformations on this Z space by means of the Gauss decomposition.

In general, let $g = (g_{ij})$ be an arbitrary invertible $n \times n$ matrix. We construct a rational transformation ρ_g on the Z space for each g. Note that the product $Z\,g$ of the upper triangular matrix Z and the general g is no longer upper triangular. So, we decompose this product in the form

$$Z\,g = X_{<0}\,X_{\geq 0}: \quad \begin{array}{l} (1)\ X_{<0} \text{ is lower unitrangular,} \\ (2)\ X_{\geq 0} \text{ is upper triangular.} \end{array} \tag{7.30}$$

Note that the minor determinants $\xi^{1\ldots r}_{1\ldots r}(Z\,g)$ $(r = 1, \ldots, n)$ of $Z\,g$ are nonzero polynomials in z_j, z_{ij}. Hence we can apply Theorem 6.3 to obtain the Gauss decomposition of $Z\,g$ as a matrix of rational functions. We then define the transformation ρ_g on the Z space by using the upper triangular component $X_{\geq 0}$ as follows:

$$\begin{aligned} \rho_g(z_j) &= \text{the } (j,j)\text{-component of } X_{\geq 0} \quad (j = 1 \ldots, n), \\ \rho_g(z_{ij}) &= \text{the } (i,j)\text{-component of } X_{\geq 0} \quad (1 \leq i < j \leq n). \end{aligned} \tag{7.31}$$

With the notation $\rho_g(Z) = (\rho_g(z_{ij}))_{ij}$, the transformation ρ_g is defined by the relation $\rho_g(Z) = X_{\geq 0}$. Since the lower triangular component also plays an important role, we denote by $G_g = X_{<0}^{-1}$ the inverse of $X_{<0}$. In this notation, the upper triangular matrix $\rho_g(Z)$ and the lower unitriangular matrix G_g are defined simultaneously by the Gauss decomposition

$$Z\,g = G_g^{-1}\,\rho_g(Z), \quad \text{namely} \quad \rho_g(Z) = G_g\,Z\,g. \tag{7.32}$$

From the uniqueness of the Gauss decomposition, it follows that, for two invertible matrices g_1, g_2, we have the product formula

$$\rho_{g_1 g_2} = \rho_{g_1}\rho_{g_2}. \tag{7.33}$$

In fact, since $\rho_{g_2}(Z) = G_{g_2} Z g_2$, we have

$$\rho_{g_1}\rho_{g_2}(Z) = \rho_{g_1}(G_{g_2})\rho_{g_1}(Z) g_2 = \rho_{g_1}(G_{g_2}) G_{g_1} Z g_1 g_2. \tag{7.34}$$

Comparing this formula with $\rho_{g_1 g_2}(Z) = G_{g_1 g_2} Z g_1 g_2$, from the uniqueness of the Gauss decomposition we obtain

$$\rho_{g_1}\rho_{g_2}(Z) = \rho_{g_1 g_2}(Z), \quad \rho_{g_1}(G_{g_2}) G_1 = G_{g_1 g_2}, \tag{7.35}$$

simultaneously. Summarizing these arguments, we have

THEOREM 7.2. *For any invertible $n \times n$ matrix g, define the rational transformation ρ_g of the variables z_j, z_{ij} and the lower unitriangular matrix G_g of rational functions through the Gauss decomposition*

$$Z\,g = G_g^{-1}\,\rho_g(Z), \quad \textit{namely} \quad \rho_g(Z) = G_g\,Z\,g. \tag{7.36}$$

Then for any invertible matrices g_1, g_2, we have

$$\begin{aligned} \rho_{g_1 g_2} &= \rho_{g_1}\rho_{g_2}, & \rho_1 &= 1, \\ G_{g_1 g_2} &= \rho_{g_1}(G_{g_2})G_{g_1}, & G_1 &= 1. \end{aligned} \tag{7.37}$$

We make use of this machinery for constructing the birational transformations of the Z space that correspond to the s_k of the M space. For each $k = 1, \ldots, n-1$, we consider the matrix

$$S_k = \begin{array}{c} \begin{matrix} & & k & k+1 & & \end{matrix} \\ \begin{bmatrix} \ddots & & & & & \\ & 1 & & & & \\ & & 0 & -1 & & \\ & & 1 & 0 & & \\ & & & & 1 & \\ & & & & & \ddots \end{bmatrix} \end{array}. \tag{7.38}$$

Note that, except for the sign, it has the same structure as the permutation matrix E_{σ_k} corresponding to the transposition. On the n-dimensional vector space $V = \mathbb{C}^n$ with canonical basis $\boldsymbol{e}_1, \ldots, \boldsymbol{e}_n$, the action of S_k is expressed as

$$S_k(\boldsymbol{e}_k) = \boldsymbol{e}_{k+1}, \quad S_k(\boldsymbol{e}_{k+1}) = -\boldsymbol{e}_k, \quad S_k(\boldsymbol{e}_j) = \boldsymbol{e}_j \quad (j \neq k, k+1). \tag{7.39}$$

Note that the minus sign appears when the index is decreased, and hence $\det S_k = 1$. (For our purpose, it is more convenient to use a matrix with determinant 1.) Although the square S_k^2 is a diagonal matrix, the k-th and the $(k+1)$-st components are -1; hence we have $S_k^2 \neq 1$ and $S_k^4 = 1$. It is easy to check that the matrices $S_1, \ldots, S_{n-1}$ satisfy the commutation relations

$$S_i^4 = 1, \quad S_i S_j = S_j S_i \quad (|i-j| \geq 2), \quad S_i S_j S_i = S_j S_i S_j \quad (|i-j| = 1). \tag{7.40}$$

We now determine the rational transformation ρ_{S_k} on the Z space induced by S_k. In the Gauss decomposition

$$Z\,S_k = X_{<0}\,X_{\geq 0}, \tag{7.41}$$

$X_{<0}$ is determined only by the block corresponding to the indices $k, k+1$. Since

$$\begin{bmatrix} z_{k,k+1} & -z_k \\ z_{k+1} & 0 \end{bmatrix} = \begin{bmatrix} 1 & 0 \\ \dfrac{z_{k+1}}{z_{k,k+1}} & 1 \end{bmatrix} \begin{bmatrix} z_{k,k+1} & -z_k \\ 0 & \dfrac{z_k\,z_{k+1}}{z_{k,k+1}} \end{bmatrix}, \tag{7.42}$$

we obtain $X_{<0} = 1 + (z_{k+1}/z_{k,k+1})E_{k+1,k}$. Hence $G_{S_k} = X_{<0}^{-1}$ is determined as

$$G_{S_k} = 1 - \frac{z_{k+1}}{z_{k,k+1}}E_{k+1,k}. \tag{7.43}$$

Furthermore, by using

$$\rho_{S_k}(Z) = G_{S_k}\,Z\,S_k, \tag{7.44}$$

we can compute the matrix $\rho_{S_k}(Z) = X_{\geq 0}$. In order to simplify the notation, we set $r_k = \rho_{S_k}$ from now on. Then the action of r_k on the z variables is determined as follows:

$$(7.45)\quad \begin{aligned} & r_k(z_k) = z_{k,k+1}, \quad r_k(z_{k+1}) = \frac{z_k z_{k+1}}{z_{k,k+1}}, \quad r_k(z_{k,k+1}) = -z_k, \\ & r_k(z_{i,k}) = z_{i,k+1}, \quad r_k(z_{i,k+1}) = -z_{i,k} \qquad (i < k), \\ & r_k(z_{k+1,j}) = z_{k+1,j} - \frac{z_{k+1} z_{k,j}}{z_{k,k+1}} \qquad (j > k+1), \\ & \text{the other } z_j,\ z_{ij}\ (i<j) \text{ are invariant.} \end{aligned}$$

We have already seen in Theorem 7.2 that the transformations ρ_g, in general, preserve the multiplicative structure of the group with respect to g. Hence, the transformations $r_1 = \rho_{S_1}, \ldots, r_{n-1} = \rho_{S_{n-1}}$ satisfy the same relations as (7.40). To summarize:

THEOREM 7.3. *Define the birational transformations r_k $(k = 1, \ldots, n-1)$ on the space of matrices Z (7.29) through the Gauss decomposition of the right action of S_k (7.38) :*

$$(7.46)\qquad Z\,S_k = G_{S_k}^{-1}\, r_k(Z), \quad \textit{namely} \quad r_k(Z) = G_{S_k}\, Z\, S_k.$$

Then, for each k, the action of r_k on the coordinate functions z_j, z_{ij} $(i<j)$ is given by the formulas (7.45). Furthermore, the birational transformations $r_1, \ldots, r_{n-1}$ satisfy the commutation relations

$$(7.47)\qquad r_i^4 = 1, \quad r_i r_j = r_j r_i \quad (|i-j| \geq 2), \quad r_i r_j r_i = r_j r_i r_j \quad (|i-j| = 1).$$

In what follows, we denote by $\mathcal{W} = \langle\, r_1, \ldots, r_{n-1} \,\rangle$ the group with fundamental relations (7.47). We have realized this group $\mathcal{W}$ as a group of birational transformations on the space of matrices Z. We remark that there is a surjective group homomorphism from $\mathcal{W} = \langle\, r_1, \ldots, r_{n-1} \,\rangle$ to the Weyl group $W = \langle\, s_1, \ldots, s_{n-1} \,\rangle$ defined by the correspondence $r_i \mapsto s_i$. Note that the relations $r_i^4 = 1$ $(i = 1, \ldots, n-1)$ are the only difference in the fundamental relations for the generators of $\mathcal{W}$, compared with those of W.

7.3. Where are the τ-functions?

Recall that, in Painlevé equations, the τ-functions play a central role. Although we are not considering differential equations in the argument of this chapter, we can introduce variables that play the same role as the τ-functions play in Bäcklund transformations. The key point is how to relate the birational transformations on the space of matrices Z of Section 7.2 to those on the space of matrices M of Section 7.1.

On the two matrices Z (7.29) and M (7.1), we impose an algebraic relation as follows: Denoting by $D(\varepsilon) = \operatorname{diag}(\varepsilon_1, \ldots, \varepsilon_n)$ the diagonal matrix consisting of the diagonal entries of M, we set

$$(7.48)\qquad M = Z\, D(\varepsilon)\, Z^{-1}.$$

Although the matrix Z is not determined uniquely from M, the ambiguity in determining Z is precisely the degree of freedom in choosing the diagonal entries

$z_1, \dots, z_n$ of Z. In order to clarify this correspondence, we introduce an intermediate matrix P that is upper unitriangular. Denoting by $D(z) = \mathrm{diag}(z_1, \dots, z_n)$ the diagonal part of Z, we set $P = ZD(z)^{-1}$:

$$P = \begin{bmatrix} 1 & p_{12} & p_{13} & \cdots & p_{1n} \\ & 1 & p_{23} & \cdots & p_{2n} \\ & & \ddots & & \vdots \\ & 0 & & 1 & p_{n-1\,n} \\ & & & & 1 \end{bmatrix}, \quad p_{ij} = \frac{z_{ij}}{z_j} \quad (i < j). \tag{7.49}$$

Then we have

$$M = ZD(\varepsilon)Z^{-1} = PD(z)D(\varepsilon)D(z)^{-1}P^{-1} = PD(\varepsilon)P^{-1}. \tag{7.50}$$

From the argument of Section 6.3, we know that M determines uniquely the sequence of eigenvalues $(\varepsilon_1, \dots, \varepsilon_n)$ and the matrix P that diagonalizes M, and *vice versa.* In other words, under the condition $\varepsilon_i \neq \varepsilon_j$ $(i \neq j)$, we have a biregular correspondence

$$(\varepsilon_j; f_{ij}) \longleftrightarrow (\varepsilon_j; p_{ij}). \tag{7.51}$$

In fact, the f_{ij} are polynomials in ε_j and p_{ij}, and conversely, the p_{ij} are polynomials in ε_j and f_{ij} except for the denominators consisting of the factors of the form $\varepsilon_{ij} = \varepsilon_i - \varepsilon_j$ $(i < j)$; we already gave the explicit formulas in (6.52) of Section 6.3.

Together with z_j, we see that the following three data are equivalent to each other:

$$(\varepsilon_j; f_{ij}; z_j) \longleftrightarrow (\varepsilon_j; p_{ij}; z_j) \longleftrightarrow (\varepsilon_j; z_{ij}; z_j). \tag{7.52}$$

Among these three coordinate systems, the subdiagonal entries are transformed into each other by

$$-\frac{f_j}{\alpha_j} = p_{jj+1} = \frac{z_{jj+1}}{z_{j+1}} \quad (j = 1, \dots, n-1). \tag{7.53}$$

In Section 7.1, we have constructed the birational transformations s_k $(k = 1, \dots, n-1)$ of the variables ε_j, f_{ij} $(i < j)$ on the M space. On the other hand, in Section 7.2, we have constructed the birational transformations r_k $(k = 1, \dots, n-1)$ of the variables z_j, z_{ij} $(i < j)$ on the Z space. Let us investigate the correspondence between these two classes of transformations under the condition that M and Z are related by the formula $M = ZD(\varepsilon)Z^{-1}$. Since the two coordinate systems

$$(\varepsilon_j; f_{ij}; z_j) \longleftrightarrow (\varepsilon_j; z_{ij}; z_j) \tag{7.54}$$

are equivalent, if we define the action of r_k on ε_j, then it induces the action of r_k on f_{ij}. We define the action of r_k on ε_j by the formula

$$r_k(D(\varepsilon)) = S_k^{-1}\, D(\varepsilon) S_k, \tag{7.55}$$

noting that the right-hand side is again a diagonal matrix. Explicitly, this definition is equivalent to

$$r_k(\varepsilon_j) = \varepsilon_{\sigma(j)} = s_k(\varepsilon_j) \qquad (j = 1, \dots, n). \tag{7.56}$$

Namely, on the ε variables the action of r_k coincides with that of s_k. Let us now determine the action of r_k on f_{ij}. Since

$$-\frac{f_j}{\alpha_j} = p_{jj+1} = \frac{z_{jj+1}}{z_{j+1}} \quad (j = 1, \dots, n-1) \tag{7.57}$$

as we have already seen in (7.53), we have

$$G_{S_k} = 1 - \frac{z_{k+1}}{z_{k,k+1}} E_{k+1,k} = 1 + \frac{\alpha_k}{f_k} E_{k+1,k} = G_k. \tag{7.58}$$

Namely, G_{S_k} is identical to the G_k that we used in Section 7.1. From this it follows that

$$\begin{aligned} r_k(M) &= r_k(ZD(\varepsilon)Z^{-1}) = r_k(Z)r_k(D(\varepsilon))r_k(Z)^{-1} \\ &= (G_k\, Z\, S_k)(S_k^{-1}D(\varepsilon)S_k)(S_k^{-1}Z^{-1}G_k^{-1}) \\ &= G_k\, Z\, D(\varepsilon)Z^{-1}G_k^{-1} = G_k M G_k^{-1}. \end{aligned} \tag{7.59}$$

This implies $r_k(M) = s_k(M)$, i.e.

$$\begin{aligned} r_k(\varepsilon_j) &= s_k(\varepsilon_j) \quad (j = 1, \dots, n); \\ r_k(f_{ij}) &= s_k(f_{ij}) \quad (1 \le i < j \le n). \end{aligned} \tag{7.60}$$

We thus conclude that the action of r_k on the variables ε_j, f_{ij} coincides with that of s_k.

Given an element $r = r_{k_1} \cdots r_{k_p} \in \mathcal{W}$, set $g = S_{k_1} \cdots S_{k_p}$. Then we have $r(Z) = G_g\, Z\, g$ and $r(D(\varepsilon)) = g^{-1}D(\varepsilon)g$. By a computation similar to the one above, we can determine the action of $r = r_{k_1} \cdots r_{k_p}$ on f_{ij} as

$$r(M) = G_g M G_g^{-1}, \quad G_g = r_{k_1} \cdots r_{k_{p-1}}(G_{k_p}) \cdots r_{k_1}(G_{k_2})G_{k_1}. \tag{7.61}$$

Since the actions of r_k on ε_j, f_{ij} are the same as those of s_k, the above computation implies that

$$w(M) = G_g M G_g^{-1}, \quad G_g = s_{k_1} \cdots s_{k_{p-1}}(G_{k_p}) \cdots s_{k_1}(G_{k_2})G_{k_1} \tag{7.62}$$

for any element $w = s_{k_1} \cdots s_{k_p} \in W$ of the Weyl group.

As for the variables ε_j and f_{ij}, we have no more than the action of W that we already had in Section 7.1; we have made a detour in that respect. But it is important to notice that through the construction of Section 7.2, we have obtained a natural extension of the action of W to the diagonal variables z_j. The action of r_k on z_j thus obtained is described as

$$\begin{aligned} r_k(z_k) &= p_{k,k+1}z_{k+1} = -\frac{f_k}{\alpha_k} z_{k+1}, \\ r_k(z_{k+1}) &= \frac{1}{p_{k,k+1}} z_k = -\frac{\alpha_k}{f_k} z_k, \\ r_k(z_j) &= z_j \qquad (j \ne k, k+1), \end{aligned} \tag{7.63}$$

in terms of f_{ij}, p_{ij}. Using these diagonal variables z_j, we can define the *τ-functions* $\tau_1 \dots, \tau_n$ as follows:

$$\tau_1 = z_1, \quad \tau_2 = z_1 z_2, \dots, \quad \tau_n = z_1 z_2 \dots z_n. \tag{7.64}$$

Conversely, the diagonal variables z_j are expressed as

$$z_1 = \tau_1, \quad z_2 = \frac{\tau_2}{\tau_1}, \dots, \quad z_n = \frac{\tau_n}{\tau_{n-1}} \tag{7.65}$$

in terms of τ-functions. From formulas (7.63) we finally obtain the following action of r_i on the τ-functions: For each $i = 1, \dots, n-1$ and $j = 1, \dots, n$,

$$r_i(\tau_j) = \tau_j \quad (i \neq j), \quad r_i(\tau_i) = p_{i,i+1} \frac{\tau_{i-1}\tau_{i+1}}{\tau_i} = -\frac{f_i}{\alpha_i} \frac{\tau_{i-1}\tau_{i+1}}{\tau_i}. \tag{7.66}$$

Notice that, except for the factor $-\alpha_i$, this action has the same structure as the Bäcklund transformations of the τ-functions for the Painlevé equation. Since $r_i(f_i) = f_i$, $r_i(\alpha_i) = -\alpha_i$, we have

$$r_i^2(\tau_i) = \frac{f_i}{\alpha_i} \frac{\tau_{i-1}\tau_{i+1}}{r_i(\tau_i)} = -\tau_i, \tag{7.67}$$

and hence

$$r_i^2(\tau_j) = \tau_j \quad (i \neq j), \quad r_i^2(\tau_i) = -\tau_i. \tag{7.68}$$

In particular, we have $r_i^4 = 1$.

By the construction discussed above, we have shown the following:

PROPOSITION 7.4. *For each $k = 1, \dots, n-1$, define the birational transformation of the variables ε_j, f_{ij} $(i < j)$, τ_j as follows:*

$$\begin{aligned} & r_k(\varepsilon_j) = \varepsilon_{\sigma_k(j)}, \\ & r_k(f_{ij}) = f_{ij} + \frac{\alpha_k}{f_k} (\delta_{k+1,i} f_{k,j} - \delta_{j,k} f_{i,k+1}) \quad (i < j), \\ & r_k(\tau_j) = \tau_j \quad (k \neq j), \quad r_k(\tau_k) = -\frac{f_k}{\alpha_k} \frac{\tau_{k-1}\tau_{k+1}}{\tau_k}. \end{aligned} \tag{7.69}$$

Then these r_k $(k = 1, \dots, n-1)$ satisfy the relations (7.47) for the generators of $\mathcal{W}$.

We remark that, for the variables f_{ij} $(i < j)$, we have $r_k^2(f_{ij}) = f_{ij}$. In fact, since S_k^2 is already diagonal,

$$r_k^2(Z) = Z S_k^2, \quad G_{S_k^2} = 1, \tag{7.70}$$

and hence $r_k^2(M) = G_{S_k^2} M G_{S_k^2}^{-1} = M$. Since we know that the action of r_k on the variables ε_j, f_{ij} $(i < j)$ coincides with that of s_k defined in Section 7.1, we have also completed the proof of Theorem 7.1.

It is also possible to define the action of s_i on the τ-functions by

$$s_i(\tau_j) = \tau_j \quad (i \neq j), \quad s_i(\tau_i) = f_i \frac{\tau_{i-1}\tau_{i+1}}{\tau_i}, \tag{7.71}$$

just as we did in the case of Painlevé equations. If we employ this definition, we have $s_i^2 = 1$ as a transformation of the variables τ_j (or z_j).

THEOREM 7.5. *For each* $k = 1, \dots, n-1$, *define the birational transformation* s_k *of the variables* ε_j, f_{ij} $(i < j)$, τ_j *as follows:*

$$
\begin{aligned}
&s_k(\varepsilon_j) = \varepsilon_{\sigma_k(j)}, \\
&s_k(f_{ij}) = f_{ij} + \frac{\alpha_k}{f_k}\left(\delta_{k+1,i} f_{k,j} - \delta_{j,k} f_{i,k+1}\right) \quad (i < j), \\
&s_k(\tau_j) = \tau_j \quad (k \neq j), \quad s_k(\tau_k) = f_k\,\frac{\tau_{k-1}\tau_{k+1}}{\tau_k}.
\end{aligned}
\tag{7.72}
$$

Then these s_k $(k = 1, \dots, n-1)$ *satisfy the relations* (7.23) *for the generators of the Weyl group* W.

In order to prove this theorem, it is sufficient to show that the relations

$$
\begin{aligned}
s_i s_j(\tau_k) &= s_j s_i(\tau_k) \qquad (|i-j| \geq 2), \\
s_i s_j s_i(\tau_k) &= s_i s_j s_i(\tau_k) \qquad (|i-j| = 1)
\end{aligned}
\tag{7.73}
$$

hold for each τ_k. Both the relations hold trivially when $k \neq i, j$. We will prove only the braid relation for $|i-j| = 1$, since the commutativity for $|i-j| \geq 2$ is easy. Setting $j = i+1$, we consider the case when $k = i, i+1$. Since $s_{i+1}s_i s_{i+1}(\tau_i) = s_{i+1}s_i(\tau_i)$, the equality $s_i s_{i+1} s_i(\tau_i) = s_{i+1}s_i s_{i+1}(\tau_i)$ is equivalent to the s_i-invariance of $s_{i+1}s_i(\tau_i)$. We show that the r_i-invariance of $r_{i+1}r_i(\tau_i)$ implies the s_i-invariance of $s_{i+1}s_i(\tau_i)$. In the expression

$$
r_{i+1}r_i(\tau_i) = r_{i+1}\left(-\frac{f_i}{\alpha_i}\frac{\tau_{i-1}\tau_{i+1}}{\tau_i}\right) = \frac{s_{i+1}(f_i)\,f_{i+1}}{(\alpha_i + \alpha_{i+1})\,\alpha_{i+1}}\,\frac{\tau_{i-1}\tau_{i+2}}{\tau_{i+1}},
\tag{7.74}
$$

the factors of τ and $(\alpha_i + \alpha_{i+1})\,\alpha_{i+1}$ are both r_i-invariant. Hence $s_{i+1}(f_i)f_{i+1}$ is r_i-invariant. Being expressed by the ε variables and the f variables, it is s_i-invariant; namely,

$$
s_i s_{i+1}(f_i) s_i(f_{i+1}) = s_{i+1}(f_i) f_{i+1}.
\tag{7.75}
$$

Since $s_{i+1}s_i(\tau_i) = s_{i+1}(f_i)f_{i+1}\tau_{i-1}\tau_{i+2}/\tau_{i+1}$, formula (7.75) implies the s_i-invariance of $s_{i+1}s_i(\tau_i)$. The case when $k = i+1$ can be treated similarly.

Note that on the variables $z_1, \dots, z_n$, the action s_i is described as

$$
s_i(z_i) = f_i\, z_{i+1}, \quad s_i(z_{i+1}) = \frac{1}{f_i}\, z_i, \quad s_i(z_j) = z_j \quad (j \neq i, i+1).
\tag{7.76}
$$

7.4. Explicit formulas of Jacobi-Trudi type

In what follows, we identify each s_i with the transposition $\sigma_i = (i, i+1)$, and W with the symmetric group $\mathfrak{S}_n$. In particular, by rewriting $w = s_{i_1} \cdots s_{i_p} \in W$ as $\sigma = \sigma_{i_1} \cdots \sigma_{i_p} \in \mathfrak{S}_n$, we set $w(j) = \sigma(j)$.

The action of s_i on $z_1, \dots, z_n$ in (7.76) can be rewritten in the form

$$
s_i(z_j) = f_i^{\epsilon}\, z_{s_i(j)}, \quad (j = 1, \dots, n),
\tag{7.77}
$$

where ϵ takes the value $+1$, -1 or 0, respectively, when $j = i$, $j = i+1$ or $j \neq i, i+1$. From this formula, it follows that, for any $w = s_{i_1} \cdots s_{i_p} \in W$, the action of w on z_j can be expressed in the form

$$
w(z_j) = f_{w,j}\, z_{w(j)},
\tag{7.78}
$$

where $f_{w,j}$ is a rational function in the ε variables and the f variables. Accordingly, for the τ-functions τ_m $(m = 1, \dots, n)$, we have

$$w(\tau_m) = w(z_1 \cdots z_m) = f_{w,1} \cdots f_{w,r}\, z_{w(1)} \cdots z_{w(m)}. \tag{7.79}$$

Given a subset $K = \{k_1, \dots, k_m\}$ $(|K| = m)$ of the indexing set $\{1, \dots, n\}$, we introduce the notation

$$z_K = \prod_{k\in K} z_k = z_{k_1} \cdots z_{k_m}. \tag{7.80}$$

Then for each $w \in W$, $w(\tau_m)$ is expressed in the form

$$w(\tau_m) = \phi_{w,m}\, z_K, \quad K = w(\{1, \dots, m\}), \tag{7.81}$$

where $\phi_{w,m} = f_{w,1} \cdots f_{w,m}$ is determined as a rational function in the ε variables and f variables. We have the same construction for $\mathcal{W}$: For each $v \in \mathcal{W}$, we obtain a rational function $\chi_{v,m}$ such that

$$v(\tau_m) = \chi_{v,m}\, z_K, \quad K = w(\{1, \dots, m\}), \tag{7.82}$$

where $w \in W$ is the canonical image of v in W.

In this section, we prove the following Jacobi-Trudi formula for the ϕ-factor $\phi_{w,m}$.

THEOREM 7.6 (Jacobi-Trudi formula of type A_{n-1}). *Given an arbitrary element w of $W = \langle s_1, \dots, s_{n-1}\rangle$ and $m \in \{1, \dots, n\}$, set $K = w(\{1, \dots, m\})$. Then the ϕ-factor $\phi_{w,m}$ depends only on the subset K. Let $k_1, \dots, k_m$ be the increasing sequence of indices obtained by arranging the elements of K. Then the transformation of the τ-function τ_m by w is given explicitly by the formula*

$$w(\tau_m) = \phi_{w,m}\, z_K = \phi_K\, z_K, \quad \phi_K = N_K \det\left(g_{i,k_j}\right)_{i,j=1}^{m}, \tag{7.83}$$

where N_K is the normalization factor

$$N_K = \prod_{k\in K, l\in K^c;\ l<k} \varepsilon_{lk}, \qquad \varepsilon_{lk} = \varepsilon_l - \varepsilon_k, \tag{7.84}$$

and the g_{ij} are defined by the determinants

$$g_{ij} = \frac{1}{\prod_{i\le k<j} \varepsilon_{kj}} \det\begin{bmatrix} f_{i\,i+1} & f_{i\,i+2} & \cdots & f_{i\,j} \\ \varepsilon_{i+1\,j} & f_{i+1\,i+2} & \cdots & f_{i+1\,j} \\ & \ddots & \ddots & \vdots \\ 0 & & \varepsilon_{j-1\,j} & f_{j-1\,j} \end{bmatrix} \tag{7.85}$$

$(g_{jj} = 1;\ g_{ij} = 0$ for $i > j)$.

Regarding the index set K as a (finite) Maya diagram, let us define the partition $\lambda = (\lambda_1, \lambda_2, \dots, \lambda_m)$ by

$$\lambda_i = k_{m+1-i} - (m + 1 - i) \qquad (i = 1, \dots, m). \tag{7.86}$$

Then we have

$$\begin{aligned} \det\left(g_{i,k_j}\right)_{i,j=1}^{m} &= \det\left(g_{m+1-i,k_{m+1-j}}\right)_{i,j=1}^{m} \\ &= \det\left(g_{m+1-i,\lambda_j+(m+1-j)}\right)_{i,j=1}^{m}. \end{aligned} \tag{7.87}$$

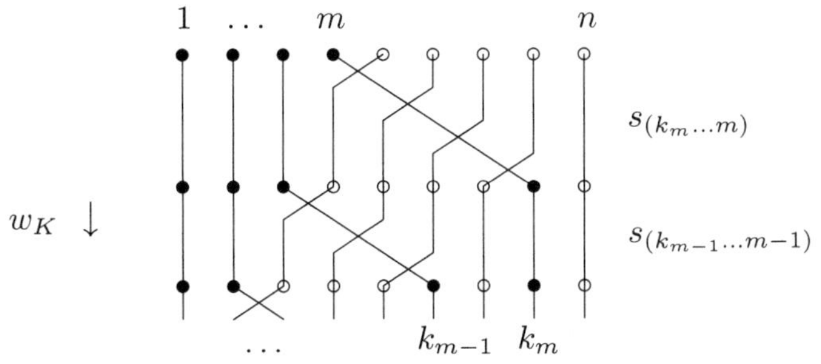

FIGURE 7.1. Decomposition of w_K

Setting $g_k^{(i)} = g_{i,i+k}$, we can rewrite this as

$$\det\left(g_{i,k_j}\right)_{i,j=1}^{m} = \det\left(g_{\lambda_j-j+i}^{(m+1-i)}\right)_{i,j=1}^{m} = \det\left(g_{\lambda_j-j+i}^{(m+1-i)}\right)_{i,j=1}^{\ell(\lambda)}. \tag{7.88}$$

In this way, we obtain a determinant of the same form as in the Jacobi-Trudi formula of Theorem 5.2. By the same argument as in the proof of Theorem 5.3, we can show that $\phi_{w,m}$ is a polynomial in α_j, f_{ij} with coefficients in $\mathbb{Z}$.

For the proof of Theorem 7.6, we first make use of the framework of $\mathcal{W}$ to determine $\chi_{v,m}$, and then derive the formula for $\phi_{w,m}$.

Let $K \subset \{1,\ldots,n\}$ be a set of indices with $|K| = m$. We arrange the elements of K and $K^{\mathrm{c}} = \{1,\ldots,n\} \setminus K$ as $k_1 < \cdots < k_m$ and $l_1 < \cdots < l_{n-m}$, respectively. Then we define the permutation $w_K \in W$ by

$$w_K = \begin{pmatrix} 1 & \ldots & m & m+1 & \ldots & n \\ k_1 & \ldots & k_m & l_1 & \ldots & l_{n-m} \end{pmatrix}. \tag{7.89}$$

LEMMA 7.7. *For each $K \subset \{1,\ldots,n\}$ with $|K| = m$, the permutation w_K has an expression $w_K = s_{i_p}s_{i_{p-1}}\ldots s_{i_1}$ satisfying the following condition: For the sequence of subsets $K_\nu = s_{i_\nu}\cdots s_{i_1}(\{1,\ldots,m\})$,*

$$i_\nu \in K_{\nu-1} \quad \text{and} \quad i_\nu + 1 \notin K_{\nu-1} \tag{7.90}$$

for each step $\nu = 1,\ldots,p$.

Using the expression

$$s_{(j\ldots i)} = s_{j-1}s_{j-2}\cdots s_i \quad : \quad \begin{pmatrix} i & i+1 & \ldots & j \\ j & i & \ldots & j-1 \end{pmatrix} \qquad (i \le j) \tag{7.91}$$

of cyclic permutation $(j\ j-1\ \ldots\ i)$, we set

$$w_K = s_{(k_1\ldots 1)}\cdots s_{(k_{m-1}\ldots m-1)}s_{(k_m\ldots m)}. \tag{7.92}$$

(See Figure 7.1.) Then this expression of w_K satisfies the condition of the lemma above.

We remark that any element $w \in W$ such that $w(\{1,\dots,m\}) = K$ can be decomposed uniquely into the product

$$w = w_K w_1 w_2; \quad w_1 \in \langle s_1,\dots,s_{m-1}\rangle, \quad w_2 \in \langle s_{m+1},\dots,s_n\rangle, \tag{7.93}$$

where w_1 is a permutation of the set $\{1,\dots,m\}$, and w_2 is a permutation of $\{m+1,\dots,n\}$. In this setting, from $w_1 w_2(\tau_m) = \tau_m$ we obtain

$$w(\tau_m) = w_K(\tau_m) = \phi_{w_K,m}\, z_K, \tag{7.94}$$

and hence $\phi_{w,m} = \phi_{w_K,m}$. This implies that the factor $\phi_{w,m}$ is determined only from the subset $K = w(\{1,\dots,m\})$. In view of this fact, we will also use the notation $\phi_{w_K,m} = \phi_K = \phi_{\{k_1,\dots,k_m\}}$. In this notation, we have $\phi_{s_{(j\dots i)},i} = \phi_{\{1\dots i-1\,j\}}$, for example.

Let us take an expression $w_K = s_{i_p}\cdots s_{i_1}$ of w_K as in Lemma 7.7, and set

$$r_K = r_{i_p}\cdots r_{i_1}, \qquad S_K = S_{i_p}\cdots S_{i_1}, \tag{7.95}$$

where S_k is the matrix defined by (7.38). Then the action of S_K on the canonical basis of $V = \mathbb{C}^n$ is given by

$$S_K(\boldsymbol{e}_j) = \boldsymbol{e}_{k_j} \qquad (j = 1,\dots,m) \tag{7.96}$$

without the minus sign. This is because each S_{i_ν} in the decomposition $S_K = S_{i_p}\cdots S_{i_1}$ *never decreases* the index of $S_{i_{\nu-1}}\cdots S_{i_1}(\boldsymbol{e}_j)$ for $j = 1,\dots,m$.

We compute $\chi_{r_K,m}$ by using the expression of $r_K = r_{i_p}\cdots r_{i_1}$ as above. Since $\tau_m = z_1\cdots z_m = \xi^{1\dots m}_{1\dots m}(Z)$, we have

$$r_K(\tau_m) = \xi^{1\dots m}_{1\dots m}(r_K(Z)) = \xi^{1\dots m}_{1\dots m}(G_{S_K}\, Z\, S_K) = \xi^{1\dots m}_{1\dots m}(Z\, S_K). \tag{7.97}$$

Here we have used the fact that $\xi^{1\dots m}_{1\dots m}$ is invariant under the left multiplication by lower unitriangular matrices. Also, from (7.96), the (i,j)-component of $Z\,S_K$ coincides with z_{i,k_j} for each $j = 1,\dots,m$. Hence we have

$$r_K(\tau_m) = \xi^{1\dots m}_{k_1\dots k_m}(Z) = \xi^{\{1\dots m\}}_K(Z). \tag{7.98}$$

We now use the matrix $P = (p_{ij})_{ij}$ of (7.49); since $Z = P\,D(z)$, we obtain

$$r_K(\tau_m) = \xi^{\{1\dots m\}}_K(Z) = \xi^{\{1\dots m\}}_K(P)\, z_K. \tag{7.99}$$

Since the entries of P are functions in the ε variables and the f variables, we see that $\chi_{r_K,m}$ is given by the determinant

$$\chi_{r_K,m} = \xi^{\{1\dots m\}}_K(P) = \det\left(p_{i,k_j}\right)_{1\le i,j\le m}. \tag{7.100}$$

Note that this formula does not depend on the choice of a decomposition $w_K = s_{i_p}\cdots s_{i_1}$ satisfying the condition of Lemma 7.7. In this sense, we can use the notation $\chi_{r_K,m} = \chi_K = \chi_{\{k_1,\dots,k_m\}}$ also for χ-factors. When the index set is $\{1,\dots,i-1,j\}$, the determinant (7.100) is precisely p_{ij}. So we can write

$$\chi_K = \det\left(p_{i,k_j}\right)_{i,j=1}^{m}, \quad p_{ij} = \chi_{\{1\dots i-1\,j\}}. \tag{7.101}$$

We remark that the ϕ-factor ϕ_K is expressed in the form

$$\phi_K = (-1)^{\ell(K)} N_K\, \chi_K, \tag{7.102}$$

by using a polynomial N_K in the ε variables, where

$$\ell(K) = \ell(w_K) = \ell(K; K^{\mathrm{c}}) = \sum_{i=1}^{m}(k_i - i). \tag{7.103}$$

The *normalization factor* N_K is determined inductively as follows, starting from $N_{\{1\dots m\}} = 1$: For each $k \in K$ such that $k+1 \notin K$,

$$N_{s_k(K)} = \alpha_k s_k(N_K). \tag{7.104}$$

In fact, when k satisfies the condition above, for the subset $L = s_k(K)$ we can take the decomposition $w_L = s_k s_{i_p} \cdots s_{i_1}$. Then from $r_L = r_k r_K$, we have

$$\chi_{r_L,m} = -\frac{f_k}{\alpha_k} s_k(\chi_{r_K,m}), \quad \phi_{w_L,m} = f_k s_k(\phi_{w_K,m}). \tag{7.105}$$

Combining this with $\ell(L) = \ell(K) + 1$, we obtain

$$N_L = (-1)^{\ell(L)} \frac{\phi_{w_L,m}}{\chi_{r_L,m}} = (-1)^{\ell(K)} \alpha_k \, s_k \left(\frac{\phi_{w_K,m}}{\chi_{r_K,m}} \right) = \alpha_k \, s_k(N_K). \tag{7.106}$$

By using the inductive structure of N_K mentioned above, we can determine the normalization factor as

$$N_K = \prod_{k \in K, \ l \in K^c; \ l<k} \varepsilon_{lk} = \prod_{k \in K, \ l \in K^c; \ l<k} (\varepsilon_l - \varepsilon_k). \tag{7.107}$$

(This formula can be proved by tracing the strings in Figure 7.1, although we will not write down the details here.) With all these arguments, we have determined the ϕ-factor $\phi_{w,m} = \phi_K$ $(K = w(\{1,\dots,m\}))$ in the general setting. When the index set is $\{1,\dots,i-1,j\}$, the normalization factor is given by $N_{\{1\dots i-1,j\}} = \prod_{i \le k<j} \varepsilon_{kj}$.

Finally, we set

$$g_{ij} = (-1)^{j-i} p_{ij}, \tag{7.108}$$

so that we obtain

$$\phi_K = N_K \, \chi_K = N_K \, \det(g_{i,k_j})_{i,j=1}^m \tag{7.109}$$

without the minus sign. The determinant expression of g_{ij} in the statement of Theorem 7.6 is nothing but the p_{ij} we discussed in Section 6.3. This completes the proof of Theorem 7.6. (In the definition of the normalization factor N_K, we excluded the sign factor $(-1)^{\ell(K)}$ so that it should become a product of *positive roots* $\varepsilon_i - \varepsilon_j$ $(i<j)$.)

Let us look at the special cases of transformation of τ_i by the cyclic permutations $s_{(j\dots i)} = s_{j-1} s_{j-2} \cdots s_i$ $(i<j)$:

$$s_{(j\dots i)}(\tau_i) = \phi_{\{1\dots i-1\,j\}} \, \tau_{i-1} z_j = \phi_{\{1\dots i-1\,j\}} \frac{\tau_{i-1}\tau_j}{\tau_{j-1}}. \tag{7.110}$$

This implies

$$\phi_{\{1\dots i-1\,j\}} = \frac{s_{(j\dots i)}(\tau_i)\tau_{j-1}}{\tau_{i-1}\tau_j} = \frac{s_{(j\dots i)}(z_i)}{z_j}. \tag{7.111}$$

Since $f_i = \phi_{\{1\dots i-1\,i+1\}}$ for $j = i+1$, the formula above is a generalization of the multiplicative formula for f_i. Note also that $\phi_{\{1\dots i-1\,j\}} = N_{\{1\dots i-1\,j\}} g_{ij}$. Hence

$\phi_{\{1\dots i-1\,j\}}$, which corresponds to the Young diagram of a single row, has the expression

$$\phi_{\{1\dots i-1\,j\}} = \det \begin{bmatrix} f_{i\,i+1} & f_{i\,i+2} & \cdots & f_{i\,j} \\ \varepsilon_{i+1\,j} & f_{i+1\,i+2} & \cdots & f_{i+1\,j} \\ & \ddots & \ddots & \vdots \\ 0 & & \varepsilon_{j-1\,j} & f_{j-1\,j} \end{bmatrix}$$
$$(7.112) \qquad = \varepsilon_{ij}\varepsilon_{i+1\,j}\cdots\varepsilon_{j-1\,j}\sum_{r=1}^{j-i}(-1)^{j-i-r}\sum_{i=i_0<i_1<\cdots<i_r=j}\frac{f_{i_0 i_1}}{\varepsilon_{i_0 j}}\cdots\frac{f_{i_{r-1} i_r}}{\varepsilon_{i_{r-1} j}}.$$

The last expansion is due to (6.53).

In this section we have described the ϕ-factor by using a subset K ($|K| = m$) of the indexing set $\{1, \dots, n\}$. This description can be translated into the language of Young diagrams, by extending K to a Maya diagram of index m with occupied boxes added on the left, and empty boxes on the right. In fact:

PROPOSITION 7.8. *There is a bijection between the set of all subsets K of $\{1, \dots, n\}$ consisting of m elements and the set of all Young diagrams contained in the rectangle $m \times (n - m)$.*

Note that each of the two sets has $\binom{n}{m}$ elements.

As we have seen already, the action $w(\tau_m) = \phi_K z_K$ of $w \in W$ on τ_m is determined by the subset $K = w(\{1, \dots, m\})$. In other words, the action of W on the subsets of $\{1, \dots, n\}$ is an indicator which governs the structure of ϕ-factors. Given a subset K, we now describe how the corresponding Young diagram changes its form according to the action of an element $w \in W = \mathfrak{S}_n$ on K. Since it is difficult to carry this out for general w, let us consider the easier case of adjacent transpositions $s_i = (i\; i+1)$ $(i = 1, \dots, n-1)$. Fixing i, we describe how the Young diagram for $s_i(K)$ is obtained from that for K. In the following two cases, K is invariant, i.e. $s_i(K) = K$:

(7.113)

$i \in K,\ i+1 \in K$:

$i \notin K,\ i+1 \notin K$:

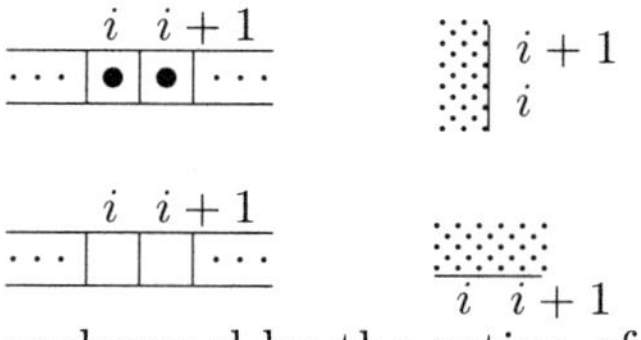

In the other two cases, i and $i+1$ are exchanged by the action of s_i:

(7.114)

$i \in K,\ i+1 \notin K$:

$i \notin K,\ i+1 \in K$:

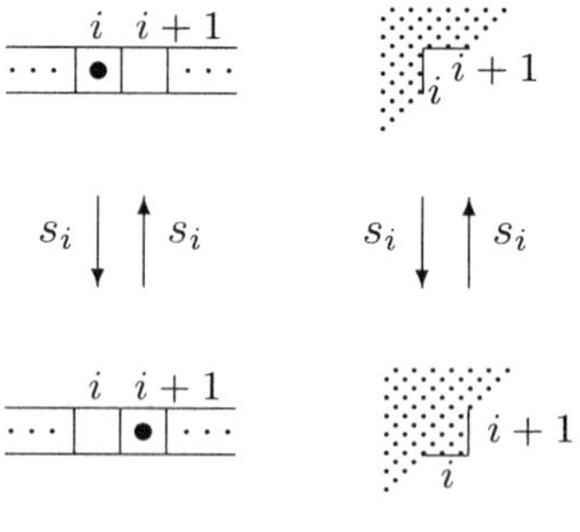

In brief, s_i does the following: Look at the two walls labeled i and $i+1$; if there is a bending, either add or remove a box there, by exchanging the roles of the horizontal and the vertical walls. Let us consider the process in which the ground state $\{1,\dots,m\}$ is transferred to $K = w(\{1,\dots,m\})$ by $w = s_{i_p}\dots s_{i_1}$. By setting $K_\nu = s_{i_\nu}\cdots s_{i_1}(\{1,\dots,m\})$, we obtain a sequence of subsets

$$\{1,\dots,m\} = K_0,\quad K_1,\quad \dots,\quad K_{p-1},\quad K_p = K \tag{7.115}$$

approaching gradually to K. In the corresponding sequence of Young diagrams

$$\emptyset = Y_0,\quad Y_1,\quad \dots,\quad Y_{p-1},\quad Y_p = Y, \tag{7.116}$$

two consecutive Young diagrams differ at most by one box. The condition

$$i_\nu \in K_{\nu-1},\quad i_\nu + 1 \notin K_{\nu-1}\quad (\nu = 1,\dots,p) \tag{7.117}$$

we have considered in Lemma 7.7 is equivalent to requiring that, at every step, the Young diagram should grow bigger by adding one box, to arrive at Y. Then, at the ν-th step, one box must be added at the corner labeled i_ν and $i_\nu+1$.

7.5. Birational transformations of type A_∞ and of type $A^{(1)}_{n-1}$

As we saw in the first half of this book, the groups of Bäcklund transformations for Painlevé equations are the affine Weyl groups of type $A^{(1)}_1$ for $P_{\rm II}$, and of type $A^{(1)}_2$ for $P_{\rm IV}$. In Section 4, we also discussed the discrete systems of type $A^{(1)}_{n-1}$ for $n \geq 3$. So far in this chapter, we have considered birational transformations of type A_{n-1} (of finite type) for arbitrary n. We will use these arguments for constructing birational transformations of type A_∞ and of type $A^{(1)}_{n-1}$. In terms of the indexing sets, the approach we are taking is first to extend the indexing set to $\mathbb{Z}$, then to *fold* it to $\mathbb{Z}/n\mathbb{Z}$:

$$\begin{array}{ccccc} A_{n-1} & \Rightarrow & A_\infty & \Rightarrow & A^{(1)}_{n-1} \\ \{1,\dots,n\} & & \mathbb{Z} & & \mathbb{Z}/n\mathbb{Z} \end{array} \tag{7.118}$$

In this way, we will obtain a further generalization of the birational transformations of type $A^{(1)}_{n-1}$ of Section 4.4; at the same time, the Jacobi-Trudi formula in this general framework will be derived naturally from the construction.

7.5.1. The case of type A_∞. This is the stage where the ϕ-factors Φ_λ that we explained at the end of Section 5.2 are going to show up. Taking $\mathbb{Z}$ for the index set, we consider the following three kinds of variables:

$$\varepsilon_j\ \ (j \in \mathbb{Z}),\quad f_{ij}\ \ (i,j \in \mathbb{Z};\ \ i<j),\quad \tau_j\ \ (j \in \mathbb{Z}). \tag{7.119}$$

Regarding these as fundamental variables, we set

$$f_j = f_{j\,j+1},\quad \alpha_j = \varepsilon_j - \varepsilon_{j+1},\quad z_j = \frac{\tau_j}{\tau_{j-1}}\qquad (j \in \mathbb{Z}). \tag{7.120}$$

Since

$$\tau_j = \tau_i\, z_{i+1} z_{i+2}\cdots z_j = \cdots z_{j-1}\, z_j, \tag{7.121}$$

we can consider each τ-function τ_j as attached to the subset $\mathbb{Z}_{\leq j}$ of $\mathbb{Z}$, i.e. the ground state of index j.

From now on, we denote the Weyl group of type A_∞ by $W(A_\infty) = \langle\, \sigma_j \ (j \in \mathbb{Z})\,\rangle$. The fundamental relations for this group are given by

$$\sigma_i^2 = 1, \quad \sigma_i\sigma_j = \sigma_j\sigma_i \ \ (|i-j| \geq 2), \quad \sigma_i\sigma_j\sigma_i = \sigma_j\sigma_i\sigma_j \ \ (|i-j| = 1). \tag{7.122}$$

By letting each σ_i correspond to the transposition $(i\ \ i+1)$, this group is identified with the group $\mathfrak{S}_\infty$ of all bijections from $\mathbb{Z}$ to itself that fix all but a finite number of elements. We denote by $\widetilde{W}(A_\infty)$ the extension of $W(A_\infty)$ obtained by adjoining an element π such that

$$\pi\sigma_i = \sigma_{i+1}\pi \qquad (i \in \mathbb{Z}). \tag{7.123}$$

When we regard $\widetilde{W}(A_\infty)$ as a group of transformations of $\mathbb{Z}$, we understand this π as representing the bijection such that $\pi(k) = k+1$ $(k \in \mathbb{Z})$.

We define the transformations σ_k $(k \in \mathbb{Z})$ and π of the variables ε_j, f_{ij} $(i<j)$, τ_j by

$$\begin{aligned}
&\sigma_k(\varepsilon_j) = \varepsilon_{\sigma_k(j)},\\
&\sigma_k(f_{ij}) = f_{ij} + \frac{\alpha_k}{f_k}\left(\delta_{k+1,i} f_{k,j} - \delta_{j,k} f_{i,k+1}\right) \quad (i<j),\\
&\sigma_k(\tau_j) = \tau_j \ \ (k \neq j), \quad \sigma_k(\tau_k) = f_k\,\frac{\tau_{k-1}\tau_{k+1}}{\tau_k},\\
&\pi(\varepsilon_j) = \varepsilon_{j+1}, \quad \pi(f_{ij}) = f_{i+1,j+1}, \quad \pi(\tau_j) = \tau_{j+1}.
\end{aligned} \tag{7.124}$$

Although we have to deal with infinitely many variables here, each element $w \in W(A_\infty)$ is relevant only to a finite number of variables among them. So one can apply the previous argument again to prove that these transformations satisfy the fundamental relations of the Weyl group of type A_∞. Also, the role of π is just to shift the indices for all the variables simultaneously. Hence we obtain a realization of the extended affine Weyl group $\widetilde{W}(A_\infty)$ as a group of birational transformations.

For given $w \in W(A_\infty)$ and $m \in \mathbb{Z}$, we set $K = w(\mathbb{Z}_{\leq m})$. Then K is a Maya diagram of index m, since w can move only a finite number of balls:

(7.125)

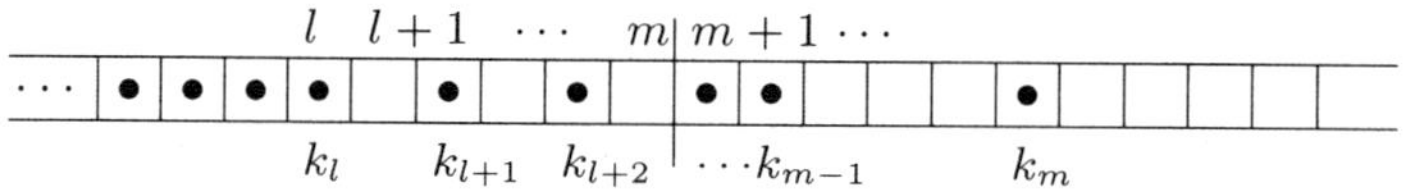

Since w acts on τ_m exactly in the same way as in the case of finite type, we see that $w(\tau_m)$ is expressed in the form

$$w(\tau_m) = \phi_K\, z_K, \quad z_K = \tau_l\, z_{k_{l+1}} \cdots z_{k_m} \quad (l \ll m). \tag{7.126}$$

Here ϕ_K is a polynomial in α_j, f_{ij}, and is determined by the determinant formula

$$\phi_K = N_K \det\left(g_{i,k_j}\right)_{i,j\leq m} = N_K \det\left(g_{i,k_j}\right)_{l<i,j\leq m} \tag{7.127}$$

of Jacobi-Trudi type. Although we are using a formal expression including a matrix of infinite size, it is essentially a determinant of a finite matrix. The normalization factor $N_K = \prod_{k\in K,\, l\in K^c;\ l<k} \varepsilon_{lk}$ is a finite product, and each g_{ij} is also defined as the determinant of a finite matrix:

$$g_{ij} = \frac{1}{\prod_{i\leq k<j}\varepsilon_{kj}} \det\begin{bmatrix} f_{i\,i+1} & f_{i\,i+2} & \cdots & f_{i\,j}\\ \varepsilon_{i+1\,j} & f_{i+1\,i+2} & \cdots & f_{i+1\,j}\\ & \ddots & \ddots & \vdots\\ 0 & & \varepsilon_{j-1\,j} & f_{j-1\,j}\end{bmatrix}. \tag{7.128}$$

Note that this determinant (apart from the denominator) is precisely the ϕ-factor attached to the Maya diagram $\{\dots, i-2, i-1, j\}$ of index i, corresponding to the Young diagram of a single row of length $j-i$.

The situation is the same when we deal with $\widetilde{w} \in \widetilde{W}(A_\infty)$, taking the shifting of indices by π into consideration. When $\widetilde{w} = \pi^r w$ ($w \in W(A_\infty)$, $r \in \mathbb{Z}$), the Maya diagram $L = \widetilde{w}(\mathbb{Z}_{\le m})$ is of index $m+r$, and the ϕ-factor for $\widetilde{w}(\tau_m)$ is given by $\phi_L = \pi^r(\phi_K)$:

$$\widetilde{w}(\tau_m) = \pi^r(\phi_K z_K) = \pi^r(\phi_K) z_{\pi^r(K)} = \phi_L\, z_L. \tag{7.129}$$

If we rewrite the right-hand side of (7.127) in terms of Young diagrams, we obtain the function $\Phi^{(m)}_\lambda$ in infinitely many variables that we introduced at the end of Section 5.2.

7.5.2. Passage to $A^{(1)}_{n-1}$. In what follows, we assume $n \ge 2$. We need two steps to move on to the case of type $A^{(1)}_{n-1}$ from the A_∞ case. The first step is to form the infinite product

$$s_k = \prod_{l\in\mathbb{Z};\ l\equiv k \pmod{n}} \sigma_l = \prod_{a\in\mathbb{Z}} \sigma_{k+an} \tag{7.130}$$

of σ_l's for each $k \in \mathbb{Z}$, where the product is taken over all $l \in \mathbb{Z}$ congruent to k modulo n. (By "$i \equiv j$", we mean "$i \equiv j \pmod{n}$".) In the setting of type A_∞, we use these s_k to construct the two groups

$$W = \langle s_0, s_1, \dots, s_{n-1} \rangle \subset \widetilde{W} = \langle s_0, s_1, \dots, s_{n-1}, \pi \rangle, \tag{7.131}$$

consisting of birational transformations acting on the ε variables, the f variables and the τ-functions. (Note that the index for s_k is periodic in the sense of $\mathbb{Z}/n\mathbb{Z}$.) The second step is to impose appropriate linear relations on the infinitely many variables ε_j, f_{ij} ($i<j$), τ_j ($i,j \in \mathbb{Z}$). As for the ε variables, we impose the relations

$$\varepsilon_j - \varepsilon_{j+n} = \delta \qquad (j \in \mathbb{Z}) \tag{7.132}$$

on them, where δ is a constant (or a new variable which is invariant under the action of $\widetilde{W}$). Under these relations, the simple roots $\alpha_j = \varepsilon_j - \varepsilon_{j+1}$ ($j \in \mathbb{Z}$) become periodic with period n, and there arises a linear relation $\alpha_0 + \cdots + \alpha_{n-1} = \delta$. As for the f variables and the τ-functions, we introduce the relations

$$f_{i+n,j+n} = f_{ij} \ \ (i<j), \qquad \tau_{j+n} = \tau_j \tag{7.133}$$

for $i, j \in \mathbb{Z}$. Then the variables $f_j = f_{j,j+1}$ satisfy the periodicity $f_{j+n} = f_j$, and the n elements $f_0, \dots, f_{n-1}$ give a set of representatives for the f_j's. (Also, the variables $z_j = \tau_j/\tau_{j-1}$ are subject to the periodicity $z_{j+n} = z_j$.)

In this procedure of passing from A_∞ to $A^{(1)}_{n-1}$, we need to do several things:

(1) We need to guarantee that the infinite products (7.130) make sense as transformations of the variables in the A_∞ setting, and to write down their actions explicitly.

(2) We need to verify that the transformations $s_0, \dots, s_{n-1}$ and π thus obtained satisfy the fundamental relations of $\widetilde{W}$ that we discussed in Section 4.4.

(3) We need to check that the relations (7.132), (7.133) to be imposed on the variables are consistent with the actions of $s_0, \dots, s_{n-1}, \pi$.

(4) We need to write down the actions of $s_0, \dots, s_{n-1}, \pi$ on the new variables of type $A_{n-1}^{(1)}$ subject to those relations.

We have almost finished the essential part of this procedure, and what remains to be done is more or less routine. So we will not write down the details of such arguments. We will describe only the resulting formulas here, and make some remarks about them.

We describe how the transformations $s_0, \dots, s_{n-1}, \pi$ act on the new variables (on which the relations (7.132), (7.133) are imposed). As for the ε variables and the τ-functions, we have

$$
(7.134)\qquad
\begin{aligned}
&s_k(\varepsilon_j) = \varepsilon_{j+1} \quad (j \equiv k), && s_k(\varepsilon_j) = \varepsilon_{j-1} \quad (j \equiv k+1),\\
&s_k(\varepsilon_j) = \varepsilon_j \quad (j \not\equiv k, k+1), &&\\
&s_k(\tau_j) = \tau_j \quad (j \not\equiv k), && s_k(\tau_j) = f_j \frac{\tau_{j-1}\tau_{j+1}}{\tau_j} \quad (j \equiv k).
\end{aligned}
$$

These formulas are the same as those in the discrete system of type $A_{n-1}^{(1)}$ discussed in Section 4.4. As for the action on f_{ij} $(i < j)$, we have

$$
(7.135)\qquad
\begin{aligned}
&s_k(f_{ij}) = f_{ij} && (i \not\equiv k+1,\ j \not\equiv k),\\
&s_k(f_{ij}) = f_{ij} + \frac{\alpha_k}{f_k} f_{i-1,j} && (i \equiv k+1,\ j \not\equiv k),\\
&s_k(f_{ij}) = f_{ij} - \frac{\alpha_k}{f_k} f_{i,j+1} && (i \not\equiv k+1,\ j \equiv k),\\
&s_k(f_{ij}) = f_{ij} + \frac{\alpha_k}{f_k}(f_{i-1,j} - f_{i,j+1}) - \left(\frac{\alpha_k}{f_k}\right)^2 f_{i-1,j+1} &&\\
& && (i \equiv k+1,\ j \equiv k).
\end{aligned}
$$

The action of π is given by

$$
(7.136)\qquad \pi(\varepsilon_j) = \varepsilon_{j+1}, \quad \pi(f_{ij}) = f_{i+1,j+1}, \quad \pi(\tau_j) = \tau_{j+1}.
$$

As we defined s_k by the infinite product (7.130), all f_{ij} with either $i \equiv k+1$ or $j \equiv k$ should be transformed simultaneously by s_k. Note that there are infinitely many such f_{ij}'s. An interesting feature arises when $i \equiv k+1$ *and* $j \equiv k$ (and hence $j \equiv i-1$). Notice that a term containing the square of α_k/f_k is present in the last formula of (7.135). We now explain how this can happen. Since $k \equiv i-1 \equiv j$, the infinite product $s_k = \prod_{l \equiv k} \sigma_l$ contains both σ_{i-1} and σ_j, which act nontrivially on f_{ij}. Since these two commute with each other, $\sigma_{i-1}\sigma_j(f_{ij})$ and $\sigma_j\sigma_{i-1}(f_{ij})$ should give the same result. Let us compute $\sigma_{i-1}\sigma_j(f_{ij})$:

$$
(7.137)\qquad
\begin{aligned}
s_k(f_{ij}) &= \sigma_{i-1}\sigma_j(f_{ij})\\
&= f_{ij} + \frac{\alpha_{i-1}}{f_{i-1}} f_{i-1,j} - \frac{\alpha_j}{f_j} f_{i,j+1} - \frac{\alpha_{i-1}\alpha_j}{f_{i-1}f_j} f_{i-1,j+1}.
\end{aligned}
$$

By using (7.133) we can rewrite this formula in terms of α_k/f_k to get the fourth formula of (7.135). In this way we naturally obtain a pattern of Bäcklund transformations like the one we already know in the $A_1^{(1)}$ case of P_{II}.

THEOREM 7.9 (Birational transformations of type $A_{n-1}^{(1)}$). *For $n \geq 2$, consider the variables ε_j, f_{ij}, τ_j $(i, j \in \mathbb{Z}; i < j)$ subject to the relations*

$$\varepsilon_{j+n} = \varepsilon_j - \delta, \quad f_{i+n,j+n} = f_{i,j}, \quad \tau_{j+n} = \tau_j. \tag{7.138}$$

Then the actions of $s_0, \ldots, s_{n-1}, \pi$ on these variables, defined by (7.134), (7.135), (7.136), give a realization of the extended affine Weyl group $\widetilde{W}$ of type $A_{n-1}^{(1)}$ as a group of birational transformations. Namely, we have $s_0^2 = 1, s_1^2 = 1$ for $n = 2$, and

$$s_j^2 = 1, \quad s_i s_j = s_j s_i \quad (j \not\equiv i \pm 1), \quad s_i s_j s_i = s_j s_i s_j \quad (j \equiv i \pm 1) \tag{7.139}$$

for $n \geq 3$. We also have $\pi s_j = s_{j+1} \pi$ for $n \geq 2$.

In this statement, the indices for s_j should be regarded as elements of $\mathbb{Z}/n\mathbb{Z}$. Note also that we have $\pi^n = 1$ when the action of $\widetilde{W}$ is restricted to the variables α_j, f_{ij}, τ_j.

REMARK 7.10. The action of $W = \langle s_0, s_1, \ldots, s_{n-1} \rangle$ on the variables f_{ij} $(i < j)$ can be expressed in terms of the Poisson bracket. In describing the Poisson bracket in the $A_{n-1}^{(1)}$ case, it is convenient to rewrite f_{ij} as

$$f_i(k) = f_{i,i+k} \quad (i \in \mathbb{Z}/n\mathbb{Z}; \quad k = 1, 2, \ldots), \tag{7.140}$$

so that $f_i = f_i(1)$. We define the Poisson bracket by

$$\{f_i(k), f_j(l)\} = \theta(j - i \equiv k)\, f_i(k + l) - \theta(i - j \equiv l)\, f_j(k + l). \tag{7.141}$$

Here $\theta(\cdot)$ stands for the function that takes the value 1 or 0 according to whether the statement is true or false. By this Poisson bracket, the action of s_i of Theorem 7.9 is written simply as

$$\begin{aligned} s_i(f_j(k)) &= f_j(k) + \frac{\alpha_i}{f_i}\{f_i, f_j(k)\} + \frac{1}{2}\left(\frac{\alpha_i}{f_i}\right)^2 \{f_i, \{f_i, f_j(k)\}\} \\ &= \exp\left(\frac{\alpha_i}{f_i}\mathrm{ad}_{\{\}}(f_i)\right)(f_j(k)). \end{aligned} \tag{7.142}$$

REMARK 7.11. In Theorem 7.9, we have used the variables of the $\mathfrak{gl}_n$-version. If one wants to deal with $\alpha_0, \ldots, \alpha_{n-1}, \delta$ and $\varepsilon_1, \ldots, \varepsilon_n, \delta$ equally, one choice is to impose another relation

$$\varepsilon_1 + \varepsilon_2 + \cdots + \varepsilon_n = 0. \tag{7.143}$$

(In this case π should be modified appropriately.) One may also wish to generate f_{ij} $(i < j)$ from f_j by the Poisson bracket. For this purpose, it would be sufficient to further assume

$$f_{i,j} + f_{i+1,j+1} + \cdots + f_{i+n-1,j+n-1} = 0 \tag{7.144}$$

for each (i, j) $(i < j)$ with $i \equiv j$, namely,

$$f_0(k) + f_1(k) + \cdots + f_{n-1}(k) = 0 \quad (k \equiv 0) \tag{7.145}$$

whenever $k > 0$ is a multiple of n. This convention is relevant to the setting of the affine Lie algebra of type $A_{n-1}^{(1)}$ in the $\mathfrak{sl}_n$-version.

7.6. Action of the affine Weyl group on the Maya diagrams

For given $w \in \widetilde{W} = \langle s_0, \ldots, s_{n-1}, \pi \rangle$ and $m \in \{0, 1, \ldots, n-1\}$, we define $K = w(\mathbb{Z}_{\le m})$. Then we obtain the ϕ-factor ϕ_K such that

$$w(\tau_m) = \phi_K \, z_K. \tag{7.146}$$

(When we work with $\widetilde{W}$, the index of K may differ from m in general.) The ϕ-factor in the case of $A_{n-1}^{(1)}$ can be determined by tracing the action of $w \in \widetilde{W} = \langle s_0, \ldots, s_{n-1}, \pi \rangle$ through $W(A_\infty)$. In order to obtain the Jacobi-Trudi formula for $A_{n-1}^{(1)}$ in the language of Maya diagrams, one has only to impose relations on α_j and f_{ij} in the Jacobi-Trudi formula for A_∞. In particular ϕ_K is determined as a polynomial in α_j, f_{ij} $(i < j)$ with coefficients in $\mathbb{Z}$. We will not repeat the description of the formula here. It would be sufficient to examine what kind of Maya diagrams (Young diagrams) should be attached to the ϕ-factors.

Let us take a Maya diagram $K \subset \mathbb{Z}$:

(7.147)
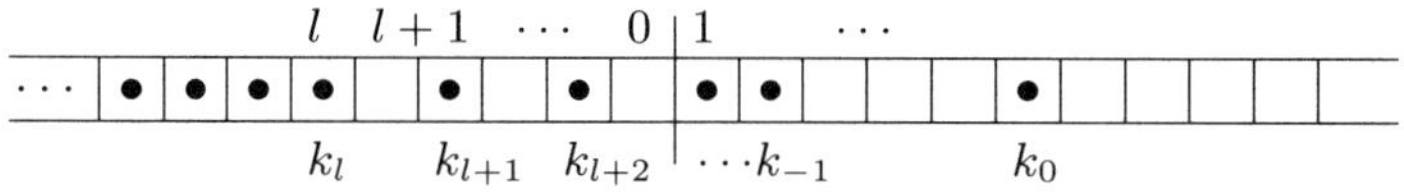

(This example has index $m = 0$.) The action of π on K results in shifting K by one to the right. Since $s_i = \prod_{j \equiv i} \sigma_j$ is an infinite product, for the action of s_i we need to consider all $j \in \mathbb{Z}$ that are congruent to i modulo n. This gives rise to a change for each pair of boxes $(j, j+1)$ $(j \equiv i)$ such that $j \in K$, $j+1 \notin K$, or $j \notin K$, $j+1 \in K$:

(7.148)
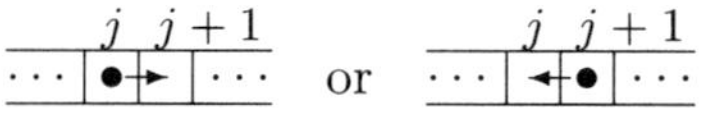

(Note that the number of such j's is always finite.) In order to visualize such a situation, we rearrange the boxes of the Maya diagram into an n-tuple of Maya diagrams. Consider the n-tuple of infinite arrays of boxes that are labeled all together as follows:

(7.149)

⋮	−7	−3	1	5	9
⋮	−6	−2	2	6	⋮
−9	−5	−1	3	7	⋮
−8	−4	0	4	8	

(This is an example with $n = 4$.) Notice that the n rows correspond to the residue classes of integers modulo n. According to this labeling, we rearrange the balls of the original Maya diagram K (7.147) into the n-tuple of arrays. As a result we obtain the following n-tuple of Maya diagrams:

(7.150)
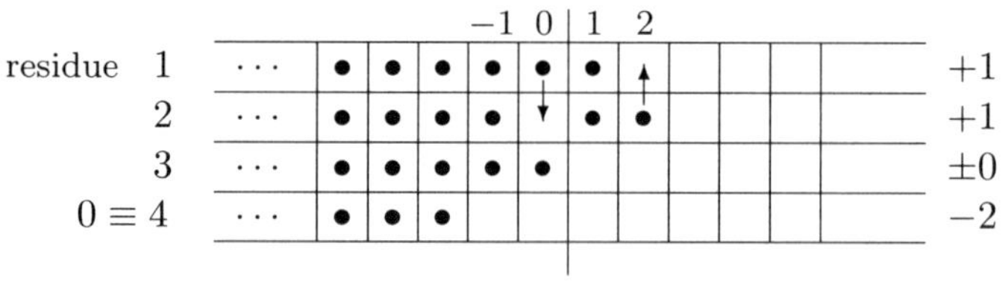

In this way, we can transform a Maya diagram into an n-tuple of Maya diagrams so that each should correspond to a residue class of labels modulo n. We now apply s_i to this picture. When $i = 1, \ldots, n-1$, the balls are exchanged between the two rows corresponding to the residues i and $i+1$. This ends up resulting in exchanging the two Maya diagrams in the i-th and $(i+1)$-th rows. As for s_0, balls are exchanged between the bottom and and the top rows; these two rows correspond to the residues 0 and 1, respectively. The action of s_0 has the following effect: Exchange first the top and the bottom rows, and then shift the top row by one to the right and the bottom row by one to the left. In the picture above, the number attached to the right of each row indicates the index of the corresponding Maya diagram, namely, the excess or deficiency as compared with $\mathbb{Z}_{\leq 0}$.

The above procedure that transforms a Maya diagram into an n-tuple of Maya diagrams corresponding to the residue classes modulo n can be reformulated as follows:

PROPOSITION 7.12. *For a given Maya diagram $K \subset \mathbb{Z}$ of index m, define an n-tuple of Maya diagrams $\boldsymbol{K} = (K_1, K_2, \ldots, K_n)$ by*

$$K_i = \{k \in \mathbb{Z}\ ;\ (k-1)n + i \in K\} \qquad (i = 1, \ldots, n). \tag{7.151}$$

If K_i has index m_i for $i = 1, \ldots, n$, then one has $m = m_1 + \ldots + m_n$. The correspondence $K \mapsto \boldsymbol{K}$ is bijective. The actions of $s_0, s_1, \ldots, s_{n-1}, \pi$ on K are then translated into the following actions on $\boldsymbol{K} = (K_1, K_2, \ldots, K_n)$:

$$\begin{aligned} s_0.\boldsymbol{K} &= (\pi(K_n), K_2, \ldots, K_{n-1}, \pi^{-1}(K_1)), \\ s_i.\boldsymbol{K} &= (K_1, \ldots, K_{i+1}, K_i, \ldots, K_n) \quad (i = 1, \ldots, n-1), \\ \pi.\boldsymbol{K} &= (\pi(K_n), K_1, \ldots, K_{n-1}). \end{aligned} \tag{7.152}$$

The action of the affine Weyl group can also be described in terms of Young diagrams. In this case we put numbers on the board as follows (we show an example of index $m = 0$):

(7.153)

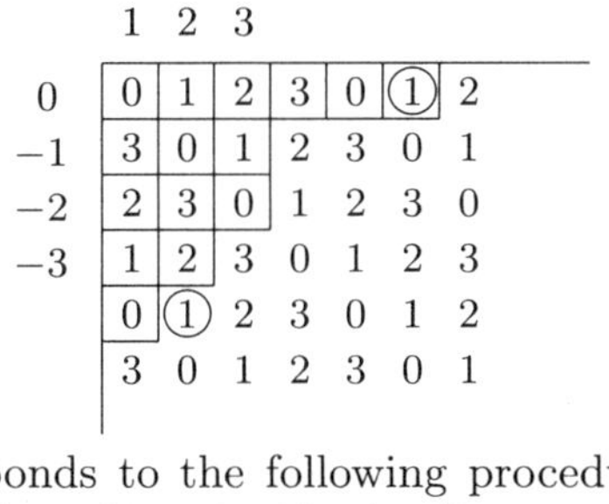

The action of s_i corresponds to the following procedure. Looking at each place numbered i, remove the box if it is inside the diagram, or add a box if it is outside the diagram, whenever one can do so keeping the condition that the resulting diagram should be a Young diagram. In the above example, by the action of s_1, the rightmost box in the first row, indicated by ①, is removed, and a box is added in the lower place also indicated by ①. (When we consider a Maya diagram of index m, we put the residue of m modulo n at the northwest corner.)

Next we examine what kind of Maya diagrams (or Young diagrams) are generated as we apply s_i $(i = 0, 1, \ldots, n-1)$ repeatedly, starting from the ground state $\mathbb{Z}_{\leq 0}$ of index 0 (the empty Young diagram). In Figure 7.2, we show how the Young diagram grows in the case of $A_2^{(1)}$; one can compare this with the situation in the case of A_∞.

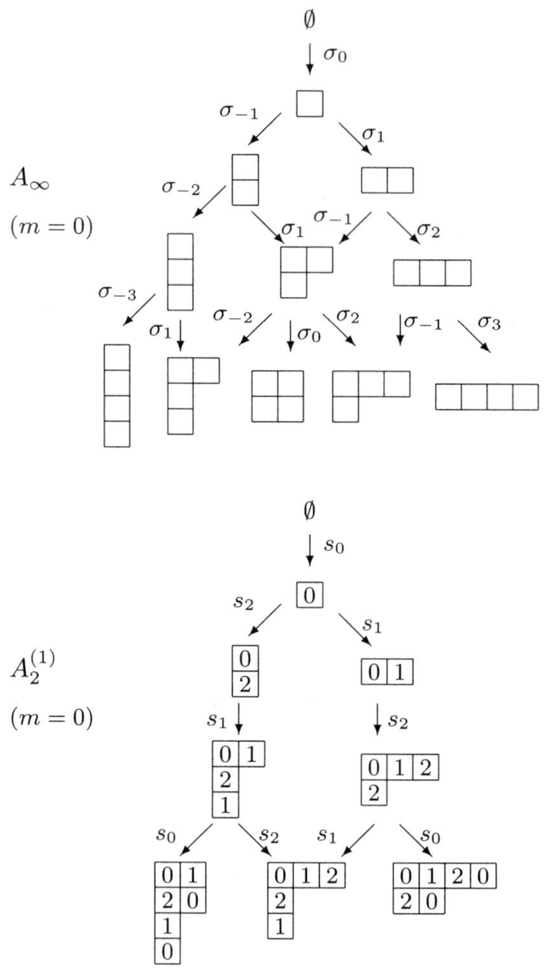

FIGURE 7.2. Generating Young diagrams

Let us consider an n-tuple of Maya diagrams as in Proposition 7.12. Note first that the ground state $\mathbb{Z}_{\leq 0}$ of index 0 corresponds to the n-tuple of ground states of index 0. At this initial stage, the only possible move is s_0:

(7.154)

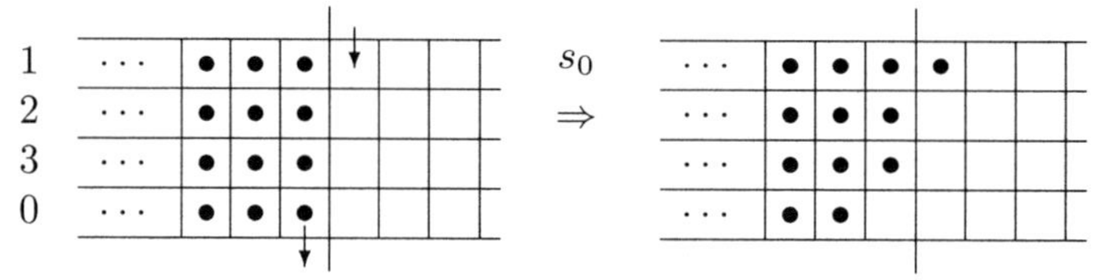

Although the indices have changed into $(1, 0, \ldots, 0, -1)$, the Maya diagram of each row remains a ground state. Since s_i only exchanges two rows (possibly with shifts), if we apply any s_i to an n-tuple of ground states, the result is again an n-tuple of ground states:

(7.155)
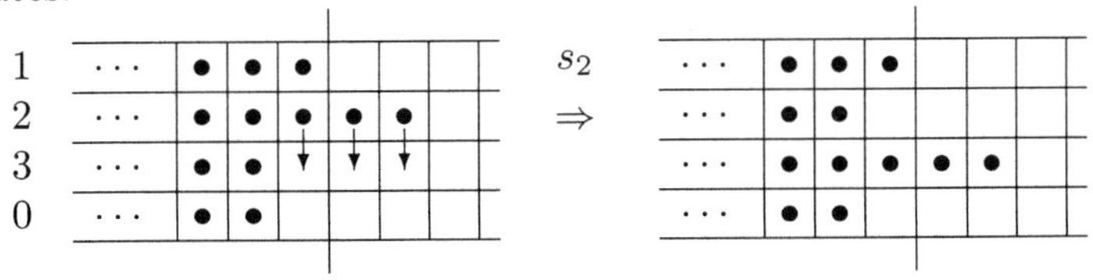

We say that a Maya diagram $K \subset \mathbb{Z}$ is an *n-core* if the rows of the corresponding n-tuple of Maya diagrams are all ground states[1]. By the action of the affine Weyl group, the n-cores are transformed again into n-cores. In particular, the Maya diagrams obtained from the ground state $\mathbb{Z}_{\leq m}$ by the action of $W = \langle s_0, \ldots, s_{n-1} \rangle$ are all n-cores of index m. Then action of π transforms an n-core of index m into an n-core of index $m+1$.

We remark that a Maya diagram $K \subset \mathbb{Z}$ is an n-core if and only if

$$k \in K, \quad l \in K^{\mathrm{C}} \text{ and } l < k \quad \Rightarrow \quad l \not\equiv k \pmod{n}. \tag{7.156}$$

Note also that the pairs (l, k) such that $k \in K$, $l \in K^{\mathrm{C}}$ and $l < k$ are in one-to-one correspondence with the boxes of the Young diagram. For such a pair (l, k), the difference $k - l$ coincides with the length of the hook that contains the corresponding box at its corner. Hence we have

PROPOSITION 7.13. *A Maya diagram $K \subset \mathbb{Z}$ is an n-core if and only if its Young diagram contains no hook whose length is a multiple of n.*

In each Young diagram appearing in Figure 7.2 of the case $A_2^{(1)}$, observe that no hook has length a multiple of 3.

PROPOSITION 7.14. *Under the action of the affine Weyl group $W = \langle s_0, \ldots, s_{n-1} \rangle$ of type $A_{n-1}^{(1)}$, the n-cores are transformed into n-cores. Furthermore, any n-core of index m is obtained from the ground state $\mathbb{Z}_{\leq m}$ by the action of some element of W.*

Exercise: Prove that for any n-core K there exist $w \in W$ and $m \in \mathbb{Z}$ such that $w(\mathbb{Z}_{\leq m}) = K$.

An n-core $K \subset \mathbb{Z}$ corresponds bijectively to an n-tuple of ground states $K_i = \mathbb{Z}_{\leq m_i}$ $(i = 1, \ldots, n)$. Hence, it can be specified by the n-tuple of indices $\boldsymbol{m} = (m_1, \ldots, m_n)$. This implies a one-to-one correspondence between the n-cores (of index m) and the n-tuples of integers $\boldsymbol{m} = (m_1, \ldots, m_n)$ (such that $m_1 + \ldots + m_n = m$). As can be clearly seen by Proposition 7.12, the actions of $s_0, s_1, \ldots, s_{n-1}$, π on the n-core K are translated into the following actions on $\boldsymbol{m}$:

$$\begin{aligned} s_0.\boldsymbol{m} &= (m_n + 1, m_2, \ldots, m_{n-1}, m_1 - 1), \\ s_i.\boldsymbol{m} &= (m_1, \ldots, m_{i+1}, m_i, \ldots, m_n) \qquad (i = 1, \ldots, n-1), \\ \pi.\boldsymbol{m} &= (m_n + 1, m_1, \ldots, m_{n-1}). \end{aligned} \tag{7.157}$$

[1] By abuse of terminology, here we call a Maya diagram (or a Young diagram) an *n-core* when its n-quotients are trivial (empty).

Notice that this is precisely the action of $\widetilde{W}$ on the lattice $L = \mathbb{Z}^n$ that we discussed in Section 4.5 in connection with the transformation of τ-functions. We now define

$$T_1 = \pi s_{n-1} \cdots s_1, \quad T_2 = s_1 \pi s_{n-1} \cdots s_2, \quad \ldots, \quad T_n = s_{n-1} \cdots s_1 \pi, \tag{7.158}$$

so that $T_i.\boldsymbol{m} = \boldsymbol{m} + \boldsymbol{e}_i$. Then for all $\nu \in L$ we have

$$T^{\nu}.\boldsymbol{m} = \boldsymbol{m} + \nu, \quad T^{\nu} = T_1^{\nu_1} \cdots T_n^{\nu}. \tag{7.159}$$

In particular we have $T^{\nu}.\mathbf{0} = \nu$.

Let us return to the transformation of the τ-functions by T^{ν}:

$$T^{\nu}(\tau_0) = \phi_K z_K, \qquad K = T^{\nu}(\mathbb{Z}_{\leq 0}). \tag{7.160}$$

The Maya diagram of this particular case is the n-core of index $|\nu| = \nu_1 + \ldots + \nu_n$ corresponding to the n-tuple of indices $\nu = (\nu_1, \ldots, \nu_n)$. This implies that its ϕ-factor is determined by the index $r = |\nu|$ and the partition $\lambda = \lambda(\nu)$. We remark that each ball in the i-th row is in correspondence with the variable $z_i = \tau_i/\tau_{i-1}$. Hence we have $z_K = \tau_0 \, z^{\nu}$. With the notation of Section 5.2, the Jacobi-Trudi formula for the ϕ-factor is given by $\phi_{T^{\nu},0} = \phi_K = \Phi_{\lambda}^{(r)}$. Hence for $r = |\nu|$ and $\lambda = \lambda(\nu)$ we obtain

$$\tau_{\nu} = T^{\nu}(\tau_0) = \Phi_{\lambda}^{(r)} \, \tau_0 \, z^{\nu} = \Phi_{\lambda}^{(r)} \, \tau_0^{1-\nu_1+\nu_n} \tau_1^{\nu_1 - \nu_2} \cdots \tau_{n-1}^{\nu_{n-1}-\nu_n}. \tag{7.161}$$

This ϕ-factor is a polynomial in α_j and f_{ij} with coefficients in $\mathbb{Z}$. Also, we see that the partitions λ that appear in this way correspond precisely to the n-cores.

In this chapter, we first constructed birational transformations for the case of the finite root system of type A_{n-1} by means of the Gauss decomposition of matrices, and showed that the Jacobi-Trudi formula for the ϕ-factors follows from the construction. Then we passed to the case of type $A_{n-1}^{(1)}$, via the case of A_{∞}, and derived a general realization of the affine Weyl group of type $A_{n-1}^{(1)}$ as a group of birational transformations, and also the Jacobi-Trudi formula for the corresponding ϕ-factors. This completes the argument on the generalization of the discrete system of type $A_{n-1}^{(1)}$ discussed in Chapters 4 and 5, as well as the determinant representation of the ϕ-factors associated with the transformations of τ-functions.

CHAPTER 8

Lax Formalism

On the basis of the Gauss decomposition of matrices, we studied in Chapters 6 and 7 the structure of birational transformations of type $A^{(1)}_{n-1}$ and ϕ-factors associated with the transformation of τ-functions. In this chapter we explain from a slightly different viewpoint how to understand the Painlevé equations and the discrete systems associated with them through the compatibility conditions of linear differential equations.

8.1. Relation with linear ordinary differential equations

The birational transformations of type $A^{(1)}_{n-1}$ constructed in Chapter 7 can be understood in connection with linear ordinary differential equations. Let us consider the following system of linear ordinary differential equations defined in a neighborhood of the origin $x = 0$ of the complex plane:

$$x\partial_x \boldsymbol{u} = A(x)\boldsymbol{u}. \tag{8.1}$$

Here $\boldsymbol{u}$ stands for a vector of n unknown functions $u_1, \ldots, u_n$, and $A(x)$ for an $n \times n$ matrix whose entries are holomorphic functions defined in a neighborhood of $x = 0$:

$$\boldsymbol{u} = \begin{bmatrix} u_1 \\ \vdots \\ u_n \end{bmatrix}, \quad A(x) = \begin{bmatrix} a_{11}(x) & \cdots & a_{1n}(x) \\ \vdots & & \vdots \\ a_{n1}(x) & \cdots & a_{nn}(x) \end{bmatrix}. \tag{8.2}$$

Also, $\partial_x = d/dx$ denotes the derivative with respect to the complex variable x. (The same notation ∂_x will be used also for $\partial/\partial x$.) The system (8.1) is a typical ordinary differential system with a *regular singular point* at $x = 0$; this particular form of singularity is sometimes called a simple singular point. We denote by $\rho_1, \ldots, \rho_n$ the eigenvalues of the matrix $A(0)$ (the constant term of $A(x)$), and assume for simplicity that $\rho_i - \rho_j \neq \mathbb{Z}$ for $i \neq j$. Then there exists an invertible matrix $P(x)$ holomorphic near $x = 0$ such that equation (8.1) is transformed into a system of *Euler equations*

$$x\partial_x \boldsymbol{v} = D(\rho)\boldsymbol{v}, \qquad D(\rho) = \mathrm{diag}(\rho_1, \ldots, \rho_n), \tag{8.3}$$

by $\boldsymbol{u} = P(x)\boldsymbol{v}$.[1] This means that the original system (8.1) has a fundamental system of solutions in the form $Y(x) = P(x)\, x^{D(\rho)}$. In this context, the eigenvalues $\rho_1, \ldots, \rho_n$ of $A(0)$ are called the *characteristic exponents* of the regular singular

[1] See, for instance, K. Iwasaki, H. Kimura, S. Shimomura and M. Yoshida, *From Gauss to Painlevé — A Modern Theory of Special Functions*, Aspects of Mathematics E16, Vieweg, 1991.

point $x = 0$. The transformation matrix $P(x)$ is determined by the system of linear differential equations

$$A(x)P(x) - P(x)D(\rho) - x\partial_x(P(x)) = 0. \tag{8.4}$$

This is equivalent to the formula

$$A(x) = P(x)D(\rho)P(x)^{-1} + x\partial_x(P(x))P(x)^{-1} \tag{8.5}$$

that transforms $D(\rho)$ into $A(x)$, and also to the relation of differential operators

$$x\partial_x - A(x) = P(x)\left(x\partial_x - D(\rho)\right)P(x)^{-1}. \tag{8.6}$$

We now assume that the matrix $A(0)$ is upper triangular, and expand $A(x)$ into a power series around $x = 0$:

$$\begin{aligned} A(x) &= A_0 + A_1 x + A_2 x^2 + \cdots \\ &= -\begin{bmatrix} \varepsilon_1 & f_{12} & \cdots & f_{1n} \\ & \varepsilon_2 & \cdots & f_{2n} \\ & & \ddots & \vdots \\ 0 & & & \varepsilon_n \end{bmatrix} - \begin{bmatrix} f_{1\,n+1} & \cdots & f_{1\,2n} \\ f_{2\,n+1} & \cdots & f_{2\,2n} \\ \vdots & & \vdots \\ f_{n\,n+1} & \cdots & f_{n\,2n} \end{bmatrix} x - \cdots \end{aligned} \tag{8.7}$$

Here we have denoted the (i,j)-component of A_k as $-f_{i,j+kn}$ $(1 \le i,j \le n)$ for $k = 0,1,\ldots$ and $i,j = 1,\ldots,n$. (Note that $\varepsilon_j = -\rho_j$ for $j = 1,\ldots,n$. This convention of sign is not essential; it is only for the sake of normalization in later arguments.) These expansion coefficients are nothing but the f variables of type $A^{(1)}_{n-1}$. Under the condition $\varepsilon_i - \varepsilon_j \notin \mathbb{Z}$ $(i \ne j)$, there exists a unique $P(x)$ such that the constant term $P(0)$ is upper unitriangular. This matrix $P(x)$ plays the role of the matrix P by which we diagonalized the matrix M.

In order to clarify the connection with the argument of Chapter 6, we consider the entries of $A(x)$ as formal power series in x. We denote by $\mathbb{C}((x))$ the field of formal Laurent series in x with at most poles at $x = 0$, and consider the infinite dimensional vector space $V = \mathbb{C}((x))^n$. By extending the canonical basis $\boldsymbol{e}_1,\ldots,\boldsymbol{e}_n$ for the n-dimensional vector space $\mathbb{Z}^n$, we define the sequence $\{\boldsymbol{e}_j\}_{j\in\mathbb{Z}}$ by

$$\ldots, x^2\boldsymbol{e}_n, x\boldsymbol{e}_1, \ldots, x\boldsymbol{e}_n, \boldsymbol{e}_1, \boldsymbol{e}_2, \ldots, \boldsymbol{e}_n, x^{-1}\boldsymbol{e}_1, \ldots. \tag{8.8}$$

(For $i = 1,\ldots,n$ and $k \in \mathbb{Z}$, we set $\boldsymbol{e}_{-kn+i} = x^k\boldsymbol{e}_i$.) Regarding the differential operator $x\partial_x - D(\rho) = x\partial_x + D(\varepsilon)$ as a linear transformation of $V = \mathbb{C}((x))^n$, we take its representation matrix with respect to the above $\{\boldsymbol{e}_j\}_{j\in\mathbb{Z}}$. Then we obtain the $\mathbb{Z}\times\mathbb{Z}$ diagonal matrix

$$D(\varepsilon_j; j \in \mathbb{Z}), \qquad \varepsilon_{j+n} = \varepsilon_j - 1 \quad (j \in \mathbb{Z}). \tag{8.9}$$

(In this version, we have $\delta = 1$.) Similarly, the representation matrix of the operator $x\partial_x - A(x)$ is given by the $\mathbb{Z}\times\mathbb{Z}$ upper triangular matrix

$$M = \left(f_{ij}\right)_{i,j\in\mathbb{Z}}; \quad f_{jj} = \varepsilon_j, \quad f_{i+n,j+n} = f_{ij} \quad (i < j). \tag{8.10}$$

If we denote the representation matrix of $P(x)$ by $P = (p_{ij})_{i,j\in\mathbb{Z}}$, then it turns out that P is a $\mathbb{Z}\times\mathbb{Z}$ unitriangular matrix and that $M = PD(\varepsilon)P^{-1}$. When we transform (8.1) into (8.3), we can also use a general invertible matrix $Z(x) = Z_0 + Z_1x + \cdots$ such that the constant term $Z(0) = Z_0$ is upper triangular. If we take the matrix representation Z of $Z(x)$, its diagonal entries are the variables z_i that are expressed as $z_i = \tau_i/\tau_{i-1}$ in terms of τ-functions.

In this setting, the action of the affine Weyl group is obtained from the transformation of dependent variables $\boldsymbol{u}$ for the system of linear equations (8.1). The transformation matrices corresponding to $s_1, \dots, s_{n-1}$ are precisely the matrices G_i that we used in Section 7.1; s_0 can also be realized by a similar matrix containing the variable x. Explicitly, we define

$$G_0(x) = 1 + \frac{\alpha_0}{f_0 x} E_{1,n}, \quad G_i(x) = 1 + \frac{\alpha_i}{f_i} E_{i+1,i} \ \ (i = 1, \dots, n-1). \tag{8.11}$$

(In fact the matrices $G_i(x)$ for $i = 1, \dots, n-1$ do not depend on x.) The rotation of indices corresponds to the matrix

$$\Lambda(x) = \sum_{i=1}^{n-1} E_{i,i+1} + E_{n,1} x = \begin{bmatrix} 0 & 1 & 0 & \cdots \\ 0 & 0 & 1 & \cdots \\ \vdots & & \ddots & \\ x & 0 & \cdots & 0 \end{bmatrix}. \tag{8.12}$$

With these matrices, we consider the following transformations of the dependent variables $u_1, \dots, u_n$ for the system of linear differential equations (8.1):

$$s_i(\boldsymbol{u}) = G_i(x)\boldsymbol{u} \ \ (i = 0, 1, \dots, n-1), \quad \pi(\boldsymbol{u}) = \Lambda(x)\boldsymbol{u}. \tag{8.13}$$

In order to write down these transformations explicitly, it is convenient to extend the indices for u_j to $j \in \mathbb{Z}$ by setting $u_{j+n} = x u_j$. Then we have $\pi(u_j) = u_{j+1}$ ($j \in \mathbb{Z}$), and for s_i ($i = 0, 1, \dots, n-1$) we have

$$s_i(u_j) = u_j + \frac{\alpha_i}{f_i} u_{j-1} \ \ (j \equiv i+1), \quad s_i(u_j) = u_j \ \ (j \not\equiv i+1) \tag{8.14}$$

for all $j \in \mathbb{Z}$. We now require that these transformations be compatible with the original system of linear equations (8.1). Namely, we require that for $w = s_0, \dots, s_{n-1}, \pi$ there should be an appropriate transformation $w(A(x))$ acting on the components of $A(x)$ (the ε variables and the f variables) such that

$$x\partial_x\, w(\boldsymbol{u}) = w(A(x))\, w(\boldsymbol{u}). \tag{8.15}$$

(We assume that w leaves x invariant.) This condition is equivalent to

$$x\partial_x - w(A(x)) = G_w(x)\,(x\partial_x - A(x))\, G_w(x)^{-1}, \tag{8.16}$$

namely

$$w(A(x))G_w(x) - G_w(x)A(x) - x\partial_x(G_w(x)) = 0. \tag{8.17}$$

One can check that this formula gives rise to the same rational transformations s_i, π for the ε variables and the f variables as we constructed in Chapter 7. In particular we see that the actions of s_i, π on the ε variables represent transformations of the characteristic exponents of the system of linear ordinary differential equations.

Let us apply our discussion of the Gauss decomposition in Chapter 7 to the present context. For each $w \in \widetilde{W}$, one can construct a transformation of the unknown functions

$$w(\boldsymbol{u}) = G_w(x)\boldsymbol{u} \tag{8.18}$$

so that the compatibility condition

$$G_{w_1 w_2}(x) = w_1(G_{w_2}(x))G_{w_1}(x) \quad (w_1, w_2 \in \widetilde{W}) \tag{8.19}$$

is satisfied. (In the construction of the previous chapter, this property was a consequence of the uniqueness of the Gauss decomposition.) In particular, for $T_1, \dots, T_n$ we obtain the matrices $G_{T_i}(x)$ such that

$$T_i(\boldsymbol{u}) = G_{T_i}(x)\boldsymbol{u}, \tag{8.20}$$

which provides a system of linear difference equations. The integrability condition

$$T_j(G_{T_i}(x))G_{T_j}(x) = T_i(G_{T_j}(x))G_{T_i}(x) \qquad (i,j = 1,\dots,n) \tag{8.21}$$

for this difference system is a special case of the compatibility condition (8.19) corresponding to $T_iT_j = T_jT_i$. The action of T_i on u_j is given explicitly by the following formula:

$$\begin{aligned} T_i(u_j) &= u_{j+1} + s_{i-1}\cdots s_{j+2}s_{j+1}\left(\frac{\alpha_j}{f_j}\right)u_j \qquad && (i > j), \\ T_i(u_i) &= u_{i+1}, && \\ T_i(u_j) &= u_{j+1} + s_{n+i-1}\cdots s_{j+2}s_{j+1}\left(\frac{\alpha_j}{f_j}\right)u_j \qquad && (i < j) \end{aligned} \tag{8.22}$$

for each $i, j = 1, \dots, n$. In terms of the τ-functions these formulas are rewritten into

$$\begin{aligned} T_i(u_j) &= u_{j+1} - (\varepsilon_i - \varepsilon_j)\frac{\tau_{j-1}T_i(\tau_j)}{\tau_jT_i(\tau_{j-1})}u_j \qquad && (i \geq j), \\ T_i(u_j) &= u_{j+1} - (\varepsilon_i - \varepsilon_j - 1)\frac{\tau_{j-1}T_i(\tau_j)}{\tau_jT_i(\tau_{j-1})}u_j \qquad && (i < j). \end{aligned} \tag{8.23}$$

We can summarize the argument above as follows. The birational transformations of type $A^{(1)}_{n-1}$ that we constructed in the previous chapter can be interpreted as the compatibility conditions of the system of linear ordinary equations coupled to the action of the affine Weyl group:

$$x\partial_x \boldsymbol{u} = A(x)\boldsymbol{u}, \quad w(\boldsymbol{u}) = G_w(x)\boldsymbol{u} \;\; (w \in \widetilde{W}). \tag{8.24}$$

In particular, the lattice part of the affine Weyl group provides the *Lax representation*

$$x\partial_x \boldsymbol{u} = A(x)\boldsymbol{u}, \quad T_i(\boldsymbol{u}) = G_{T_i}(x)\boldsymbol{u} \;\; (i = 1,\dots,n) \tag{8.25}$$

of the corresponding difference system. The general discrete system of type $A^{(1)}_{n-1}$ can be viewed as an extension of such a difference system to the affine Weyl group action.

Remark: The idea of the Gauss decomposition could have been developed in this setting of ordinary differential equations; that way might have been more transparent than the argument that we actually adopted in the previous chapters. To be more precise, take a fundamental system of solutions $Y(x) = Z(x)x^{-D(\varepsilon)}$ around $x = 0$, and multiply it by the matrix corresponding to r_i from the right. Then by applying the Gauss decomposition (the Riemann-Hilbert decomposition) to the matrix, we obtain the gauge transformation $G_i(x)$ that gives rise to a system of linear differential equations of the same form as the original one. In the construction of Chapter 7, we avoided this direct approach, and employed instead an algebraic construction by way of an intermediate step of type A_∞. A reason for this is that in this book I do not wish to use the Riemann-Hilbert decomposition. In any case, it is worth mentioning here that the expansion coefficients of the fundamental system of solutions around $x = 0$ are expressed in terms of the Bäcklund transformations of the τ-functions. Also, the ϕ-factors provide a mechanism for determining the expansion coefficients of the fundamental system of solutions from the coefficients of the system of linear differential equations.

8.2. Symmetric forms of $P_{\rm II}$, $P_{\rm IV}$ and the Lax representations

One origin of the Painlevé equations is the so-called monodromy-preserving deformations. This means that each of the Painlevé equations arises as the compatibility condition for a certain deformation of a linear differential equation. In this section we explain a way to derive the symmetric forms of $P_{\rm II}$ and $P_{\rm IV}$ from the compatibility condition of a system of linear differential equations of the form

$$x\partial_x \boldsymbol{u} = A(x)\boldsymbol{u}, \quad \partial_t \boldsymbol{u} = B(x)\boldsymbol{u} \tag{8.26}$$

with respect to the variables (x, t). Here $A(x)$ and $B(x)$ are matrices depending on (x, t), although we have suppressed their dependence on t in the notation. A standard (traditional) way to derive a Painlevé equation is to consider the compatibility condition for a single differential equation of second order and its deformation equation, but we employ here a system of linear differential equations instead. (For $P_{\rm IV}$, we will use below a linear differential system of 3×3 matrices. This derivation is slightly different from the usual one, in which either a single linear equation of second order or a 2×2 system is used.) Although the linear differential system with respect to the x-variable has the same form as (8.1) in the previous section, we consider here the case where $A(x)$ is a polynomial in x. In particular, we are dealing with a deformation of a linear differential system with a regular singular point at $x = 0$ and an irregular singular point at $x = \infty$. In this context the Bäcklund transformations (the structure of the discrete system) are relevant to the regular singular point of the linear system at $x = 0$.

In order to clarify the idea, we first make some remarks on the *compatibility condition*. If the equation (8.26) has a fundamental system of solutions, we must have

$$\partial_t \, x\partial_x \boldsymbol{u} = x\partial_x \, \partial_t \boldsymbol{u}. \tag{8.27}$$

The left-hand side is computed as

$$\begin{aligned} \partial_t \, x\partial_x \boldsymbol{u} &= \partial_t\big(A(x)\boldsymbol{u}\big) = \partial_t\big(A(x)\big)\boldsymbol{u} + A(x)\partial_t \boldsymbol{u} \\ &= \big(\partial_t\big(A(x)\big) + A(x)B(x)\big)\,\boldsymbol{u}, \end{aligned} \tag{8.28}$$

while the right-hand side is given by

$$\begin{aligned} x\partial_x \, \partial_t \boldsymbol{u} &= x\partial_x\big(B(x)\boldsymbol{u}\big) = x\partial_x\big(B(x)\big)\boldsymbol{u} + B(x)x\partial_x \boldsymbol{u} \\ &= \big(x\partial_x\big(B(x)\big) + B(x)A(x)\big)\,\boldsymbol{u}. \end{aligned} \tag{8.29}$$

Hence we have

$$\big(\partial_t\big(A(x)\big) - x\partial_x\big(B(x)\big) + [A(x), B(x)]\big)\,\boldsymbol{u} = 0. \tag{8.30}$$

Since this equation is satisfied for a general solution $\boldsymbol{u}$, we should have

$$\partial_t\big(A(x)\big) - x\partial_x\big(B(x)\big) + [A(x), B(x)] = 0. \tag{8.31}$$

This is the compatibility condition we are going to discuss. In terms of the differential operators, this condition is equivalently written as

$$[x\partial_x - A(x), \partial_t - B(x)] = 0. \tag{8.32}$$

(If you feel this argument is unsatisfactory, replace $\boldsymbol{u}$ by a fundamental system of solutions $Y = Y(x,t)$. If Y is an $n \times n$ matrix satisfying (8.26) and if Y is invertible at each $x \neq 0$, then (8.31) follows from (8.30).)

8.2.1. The case of P_{II}. In this case we take the following 2×2 matrices $A(x)$ and $B(x)$ $(n = 2)$:

$$\begin{aligned} -A(x) &= \begin{bmatrix} \varepsilon_1 & f_1 \\ 0 & \varepsilon_2 \end{bmatrix} + \begin{bmatrix} g & -2 \\ f_0 & -g \end{bmatrix} x + \begin{bmatrix} 0 & 0 \\ -2 & 0 \end{bmatrix} x^2, \\ B(x) &= \begin{bmatrix} q & -1 \\ 0 & -q \end{bmatrix} + \begin{bmatrix} 0 & 0 \\ -1 & 0 \end{bmatrix} x. \end{aligned} \tag{8.33}$$

Here the variables appearing as matrix coefficients are functions depending only on t. Although -2 and -1 may be replaced by any nonzero constants, we have chosen them for the sake of normalization. The compatibility condition (8.31) is a system of nonlinear equations with respect to the t-variables for the matrix coefficients of $A(x)$ and $B(x)$. In this case it is written down explicitly as

$$\begin{aligned} &\varepsilon_1' = \varepsilon_2' = 0, \quad g = 2q, \quad g' = f_1 - f_0, \\ &f_0' = -2qf_0 + 1 - \varepsilon_1 + \varepsilon_2, \quad f_1' = 2qf_1 + \varepsilon_1 - \varepsilon_2, \end{aligned} \tag{8.34}$$

where $' = \partial_t$. This implies in particular that the characteristic exponents $-\varepsilon_1$ and $-\varepsilon_2$ do not depend on t, and that the monodromy is invariant. By eliminating g, we see that the remaining part is precisely the symmetric form of P_{II} for f_0 and f_1.

For these matrices $A(x)$ and $B(x)$, let us consider the following system of linear equations coupled to gauge transformations:

$$\begin{aligned} x\partial_x \boldsymbol{u} &= A(x)\boldsymbol{u}, \quad & \partial_t \boldsymbol{u} &= B(x)\boldsymbol{u}, \\ s_i(\boldsymbol{u}) &= G_i(x)\boldsymbol{u}, \quad & \pi(\boldsymbol{u}) &= \Lambda(x)\boldsymbol{u}. \end{aligned} \tag{8.35}$$

The Bäcklund transformations for the symmetric form of P_{II} are obtained from the compatibility conditions for this system. We remark that the Bäcklund transformations for P_{II} can be obtained as a specialization of the general discrete system of type $A_1^{(1)}$. From the matrix $A(x)$ we can read off how the variables f_{ij} should be specialized in this setting. (Please check that this specialization coincides with the one we used in the description of the Jacobi-Trudi formula for the ϕ-factors.)

8.2.2. The case of P_{IV}. In this case we take the following 3×3 matrices $A(x)$, $B(x)$ $(n = 3)$:

$$\begin{aligned} -A(x) &= \begin{bmatrix} \varepsilon_1 & f_1 & 1 \\ 0 & \varepsilon_2 & f_2 \\ 0 & 0 & \varepsilon_3 \end{bmatrix} + \begin{bmatrix} 0 & 0 & 0 \\ 1 & 0 & 0 \\ f_0 & 1 & 0 \end{bmatrix} x, \\ B(x) &= \begin{bmatrix} q_1 & -1 & 0 \\ 0 & q_2 & -1 \\ 0 & 0 & q_3 \end{bmatrix} + \begin{bmatrix} 0 & 0 & 0 \\ 0 & 0 & 0 \\ -1 & 0 & 0 \end{bmatrix} x. \end{aligned} \tag{8.36}$$

The compatibility condition for this case is written down explicitly as

$$\begin{aligned}
&\varepsilon_1' = \varepsilon_2' = \varepsilon_3' = 0, \\
&f_1 - f_2 = -q_1 + q_3, \qquad f_0' = f_0(-q_1 + q_3) + (1 - \varepsilon_1 + \varepsilon_3), \\
&f_2 - f_0 = q_1 - q_2, \qquad f_1' = f_1(q_1 - q_2) + (\varepsilon_1 - \varepsilon_2), \\
&f_0 - f_1 = q_2 - q_3 \qquad f_2' = f_2(q_2 - q_3) + (\varepsilon_2 - \varepsilon_3).
\end{aligned} \tag{8.37}$$

The variables q_1, q_2, q_3 can be eliminated from these formulas. As a result we obtain the symmetric form of $P_{\rm IV}$ for the variables f_0, f_1, f_2.

In the same way as in the previous case, the Bäcklund transformations for $P_{\rm IV}$ are then described by the *simple version* of the discrete system of type $A_2^{(1)}$ obtained by specializing the variables f_{ij} in accordance with $A(x)$.

8.2.3. The general case. The construction of Lax forms explained above can be extended to higher order linear equations. From them we obtain a class of nonlinear differential equations whose Bäcklund transformations are described by the discrete system of type $A_{n-1}^{(1)}$. Generalizing the case of $P_{\rm IV}$, for $n \geq 3$ let us consider the following $n \times n$ matrices:

$$\begin{aligned}
-A(x) &= \mathrm{diag}(\varepsilon_1, \ldots, \varepsilon_n) + \mathrm{diag}(f_1, \ldots, f_{n-1}, f_0)\Lambda(x) + \Lambda(x)^2, \\
B(x) &= \mathrm{diag}(q_1, \ldots, q_n) + \Lambda(x).
\end{aligned} \tag{8.38}$$

(In the following we will not care about the normalization.) The nonlinear ordinary differential equation obtained from the compatibility condition of this case has the discrete system of type $A_{n-1}^{(1)}$ corresponding to the matrix $A(x)$ as its Bäcklund transformations. When $n = 4$, the corresponding nonlinear equation is essentially of second order, and is in fact equivalent to the Painlevé equation $P_{\rm V}$. When $n = 2m + 1$ (resp. $n = 2m + 2$), we obtain a higher order nonlinear equation of order $2m$ that generalizes the symmetric form of $P_{\rm IV}$ (resp. of $P_{\rm V}$, which we referred to at the end of Chapter 4).

So far we have discussed only nonlinear ordinary differential equations. More generally, we can construct a class of nonlinear partial differential equations which have the discrete systems of type $A_{n-1}^{(1)}$ as their Bäcklund transformations. Fix a natural number $r \geq 2$, and suppose r is not a multiple of n. Then we take an $n \times n$ matrix $A(x)$ of the form

$$-A(x) = \mathrm{diag}(\varepsilon_1, \ldots, \varepsilon_n) \tag{8.39}$$

$$+ \sum_{k=1}^{r-1} \mathrm{diag}(f_{1,1+k}, \ldots, f_{n,n+k})\Lambda(x)^k + \Lambda(x)^r. \tag{8.40}$$

Introducing the set of degrees $E = \{m \in \mathbb{Z}; 1 \leq k \leq r, k \not\equiv n\}$, for each $m \in E$ we take a matrix

$$B_m(x) = \sum_{k=0}^{m-1} \mathrm{diag}(q_{1,1+k}^{(m)}, \ldots, q_{n,n+k}^{(m)})\Lambda(x)^k + \Lambda(x)^m. \tag{8.41}$$

Then we consider the following system of linear differential equations with respect to the variables x and t_m ($m \in E$):

$$x\partial_x \boldsymbol{u} = A(x)\boldsymbol{u}, \quad \partial_{t_m} \boldsymbol{u} = B_m(x)\boldsymbol{u} \ \ (m \in E). \tag{8.42}$$

Then we obtain a system of nonlinear partial differential equations from the compatibility condition between the linear equation for the x variable and those for the t variables

$$\partial_{t_m}(A(x)) - x\partial_x(B_m(x)) + [A(x), B_m(x)] = 0 \quad (m \in E), \tag{8.43}$$

together with the compatibility condition among the linear equations for the t variables

$$\partial_{t_m}(B_l(x)) - \partial_{t_l}(B_m(x)) + [B_l(x), B_m(x)] = 0 \quad (l, m \in E). \tag{8.44}$$

The compatibility condition (8.44) for the matrices $B_m(x)$ is the Zakharov-Shabat equations for the n-reduced *modified KP hierarchy.* (It is a variation of the so-called Drinfeld-Sokolov hierarchy associated with the affine Lie algebra of type $A_{n-1}^{(1)}$.) The compatibility condition (8.43) between $A(x)$ and $B_m(x)$ then corresponds to imposing a certain homogeneity constraint (the similarity reduction) on the n-reduced modified KP hierarchy. The system of nonlinear partial differential equations thus obtained admits the affine Weyl group of type $A_{n-1}^{(1)}$ as a group of Bäcklund transformations in the same way as in the cases of P_{II}, P_{IV} and P_{V}. In this sense, the construction described above can be thought of as a generalization of Painlevé equations to many variables. As we have already discussed, the discrete system of type $A_{n-1}^{(1)}$ has a class of special solutions expressed in terms of Schur functions. We remark that this property of the discrete system of type $A_{n-1}^{(1)}$ is relevant to the fact that the Schur functions attached to the n-cores give special solutions to the modified KP hierarchy as well.

Appendix

A.0. Profile of Painlevé equations

By the *Painlevé equations* we mean a class of second order nonlinear ordinary differential equations discovered by Paul Painlevé about one hundred years ago. They are traditionally classified into the six equations denoted by P_{I}, P_{II}, ..., P_{VI} (see Table 0.1).

For a linear differential equation, the location of singularities appearing in the solutions can be specified by the coefficients of the equation. This is one of the characteristic features of linear differential equations. In the case of a nonlinear equation, however, the situation is quite different. We encounter the phenomenon that the singularities arise arbitrarily at different places depending on the individual solutions. Roughly speaking, a nonlinear equation is said to have the *Painlevé property* if all such *movable singularities* of the solutions are as simple as poles (or branching points of finite degree, depending on the situation). Ever since Sofya Kovalevskaya (before Painlevé) paid attention to this property in her studies on the dynamics of tops, the Painlevé property has been employed as a criterion of a *good* (or an integrable) differential equation. (The theory of infinite integrable systems has greatly developed during the latter half of the twentieth century. In many aspects of the development, the Painlevé property played the role of a practical criterion for the integrability of differential equations. I am not sure, however, whether it is completely understood why the Painlevé property should imply integrability.)

Several years after Kovalevskaya's tops, Painlevé tried to classify all the *rational* ordinary differential equations of second order with the Painlevé property, assuming that y'' should be expressed as a rational function (a ratio of polynomials) of y, y' and t. (It must have been hard work, as we can imagine.) In the end he was able to conclude that

> Any rational ordinary differential equation of second order with the Painlevé property is reduced to one of the six equations $P_{\mathrm{I}}, \ldots, P_{\mathrm{VI}}$ unless it can be integrated algebraically, or transformed into a simpler equation such as a linear equation or the differential equation of elliptic functions.

(I should also mention here the name of B. Gambier, who was a student of Painlevé and contributed to this classification.)

As a background of this classification, Painlevé had a strong motivation to find a "new class of special functions" defined by nonlinear differential equations. Probably, this was one of the main motifs of mathematics around the turn of the twentieth century. With the expectation that they should be so, the solutions of the Painlevé equations were called the *Painlevé transcendents*. Around the period of discovery, there was a controversy about whether the solutions of the Painlevé equations are *really* transcendental, although it did not result in any reasonable conclusion.

Soon after the discovery of the Painlevé equations, the relationship was already recognized between these nonlinear equations and the deformation of second order linear differential equations. In particular, P_{VI} is interpreted as the differential equation to be satisfied by the apparent singularity of a second order Fuchsian equation with four regular singular points on $\mathbb{P}^1$ as it deforms so that the monodromy is preserved (R. Fuchs). The other Painlevé equations can also be understood through the confluence of such linear equations. Around the same time, several attempts were made to generalize them to higher orders and to many variables (J. Chazy, R. Garnier), but that period of time was probably too early for such goals. In the late 1910's, the Painlevé equations withdrew from the stage of "modern mathematics".

It was really a "bolt from the blue"[2] when the Painlevé equations revived in the 1970's after a long slumber of sixty years. In the work of T.T. Wu and others on the 2-dimensional Ising model, it was revealed that P_{III} (with special parameters) appears in the scaling limit of correlation functions. The research environment on Painlevé equations seems to have changed drastically since then. A series of works by Okamoto opened a new epoch in the study of Painlevé equations. Also, a new tide toward infinite integrable systems was generated through holonomic quantum field theory (Sato, Miwa, Jimbo) and the general theory of monodromy-preserving deformations (Jimbo, Miwa, Ueno). It is not by chance that the τ-functions play the fundamental role both in Painlevé equations and in infinite integrable systems. Umemura's work in the mid-1980's brought a historical solution to the problem of irreducibility (transcendence) of Painlevé equations, and gave an impetus for the study of Painlevé equations to expand in various directions. Also, Aoki, Kawai and Takei started a new approach to Painlevé transcendents from the viewpoint of singular perturbation. There has been enormous progress in the discretization of Painlevé equations and their connection with the algebraic geometry of rational surfaces. It is probably too early to give an overview of the study of Painlevé equations during the last quarter of the twentieth century, but the mathematics related to Painlevé equations seems to have already spread beyond the scope of a survey.

Symmetry of Painlevé equations

Each of the six Painlevé equations has its own characteristics, while all being intimately related with each other (Table A.1). Here we give some remarks concerning the symmetry of Painlevé equations.

A series of "Studies on the Painlevé equations" by K. Okamoto, started in the late 1970's, has particular importance as it provides us in many respects with starting points that lead to recent studies on Painlevé equations. One thing to note is that he introduced the concept of τ-functions for Painlevé equations on the basis of the fact that each of the six Painlevé equations can be written as a Hamiltonian system. By analyzing the differential equation to be satisfied by the τ-functions, he discovered that:

---

[2] K. Okamoto, *Introduction to Painlevé Equations*, Sophia Kokyuroku in Mathematics No. 19, 1985 (in Japanese).

TABLE A.1. Profile of Painlevé Equations

Painlevé Equation	Partition of 4	Number of Parameters	Hypergeometric Solutions	Bäcklund Symmetry	Space of Solutions
P_{I}	·	0	·	·	$E_8^{(1)}$
P_{II}	4	1	Airy	$A_1^{(1)}$	$E_7^{(1)}$
P_{III}	2+2	2	Bessel	$A_1^{(1)} \oplus A_1^{(1)}$	$D_6^{(1)}$
P_{IV}	3+1	2	Hermite	$A_2^{(1)}$	$E_6^{(1)}$
P_{V}	2+1+1	3	Kummer	$A_3^{(1)}$	$D_5^{(1)}$
P_{VI}	1+1+1+1	4	Gauss	$D_4^{(1)}$	$D_4^{(1)}$

TABLE A.2. The Hypergeometric Equation and its Confluences

$$\begin{aligned}
\text{Airy:} \quad & u'' - t\,u = 0 \\
\text{Bessel:} \quad & u'' + \frac{1}{t}\,u' + \left(1 - \frac{\alpha^2}{t^2}\right) u = 0 \\
\text{Hermite:} \quad & u'' - 2\,t\,u' + 2\,\alpha\,u = 0 \\
\text{Kummer:} \quad & u'' + \left(\frac{\gamma}{t} - 1\right) u' - \frac{\alpha}{t}\,u = 0 \\
\text{Gauss:} \quad & u'' + \frac{\gamma - (\alpha+\beta+1)t}{t(1-t)}\,u' - \frac{\alpha\beta}{t(1-t)}\,u = 0
\end{aligned}$$

For each $J = \mathrm{II}, \mathrm{III}, \ldots, \mathrm{VI}$, the parameter space for P_J can be identified with the Cartan subalgebra of a semisimple Lie algebra. Furthermore, the natural action of the corresponding affine Weyl group on the Cartan subalgebra lifts to the level of dependent variables of the Painlevé equation, to form a group of *Bäcklund transformations* (birational canonical transformations).

In view of this point, we realize that P_J has *invariant divisors* (according to Umemura's terminology) along reflection hyperplanes (mirrors) of the affine Weyl group in the parameter space. By specializing the equation to each invariant divisor, we obtain a Riccati equation, from which we obtain a 1-parameter family of classical solutions of hypergeometric type. (The corresponding hypergeometric equations are indicated in Table A.2.) It also turns out that Painlevé equations have no essential invariant divisors other than those appearing along reflection hyperplanes. From this it follows that if the parameters are generic, the Painlevé equation has no transcendental classical solution. This argument is the essence of Umemura's theory on irreducibility.

When an automorphism of the Dynkin diagram lifts to a Bäcklund transformation, there arise algebraic solutions (including rational solutions) at its fixed points.

The τ-functions corresponding to such algebraic solutions and their Bäcklund transformations are expressed in terms of certain *special polynomials* (Yablonski-Vorobiev, Okamoto and Umemura polynomials). It is also observed that those special polynomials have remarkable combinatorial properties.

Recently, Yasuhiko Yamada and I found that the Bäcklund transformations of Painlevé equations can be described in a systematic way in terms of root systems. By abstraction, we also obtained a method to realize the Weyl group associated with an arbitrary root system (GCM) as a group of birational canonical transformations. In the cases of affine root systems, this class of representations describes the symmetry structure of nonlinear (ordinary or partial) differential equations of Painlevé type obtained by a certain similarity reduction from the Drinfeld-Sokolov hierarchy. (In fact, $P_{\rm II}$, $P_{\rm IV}$, $P_{\rm V}$ can be understood as arising from such infinite integrable hierarchies.) It also provides a method to construct a class of discrete dynamical systems (discrete Painlevé equations) associated with general affine Weyl groups.

This book is meant to be an introduction to such aspects of the affine Weyl group symmetry of Painlevé equations, based on the explicit computation of the two examples $P_{\rm II}$ and $P_{\rm IV}$.

A.1. Hamiltonian systems [Section 1.1]

A *Hamiltonian system* is a system of ordinary differential equations for the dependent variables (q, p) expressed in the form

$$\frac{dq}{dt} = \frac{\partial H}{\partial p}, \quad \frac{dp}{dt} = -\frac{\partial H}{\partial q}, \tag{A.1}$$

where $H = H(q, p; t)$ is a function of the variables (q, p, t), called the *Hamiltonian*. For an arbitrary given solution $q = q(t)$, $p = p(t)$, we define

$$h(t) = H(q(t), p(t); t) \tag{A.2}$$

by substituting the solution into the Hamiltonian. Let us compute the t-derivative dh/dt of the function $h(t)$ thus obtained:

$$\begin{aligned} \frac{dh}{dt} &= \frac{\partial H}{\partial q}\frac{dq}{dt} + \frac{\partial H}{\partial p}\frac{dp}{dt} + \frac{\partial H}{\partial t} \\ &= \frac{\partial H}{\partial q}\frac{\partial H}{\partial p} - \frac{\partial H}{\partial p}\frac{\partial H}{\partial q} + \frac{\partial H}{\partial t} = \frac{\partial H}{\partial t}. \end{aligned} \tag{A.3}$$

Here $\partial/\partial t$ stands for the partial derivative with (q, p, t) regarded as independent variables; when we use d/dt, we regard q and p as functions of t. (We also use the notation dH/dt instead of dh/dt, unless misunderstanding might occur.) From this computation, we see that the function $h(t)$ is a *constant* if $H = H(q, p; t)$ does not depend on t explicitly. In this case, the Hamiltonian H provides a first integral of the system (A.1). Once you find a first integral, in a generic situation, p can be thought of as a function of q that is defined implicitly by $H = c$. Hence, the original system reduces to an equation of first order for q; the unknown function q should be determined by quadrature.

Question: Write down the Hamiltonian system defined by the Hamiltonian $H = \frac{1}{2}p^2 - 2q^3 - aq$ $(a \in \mathbb{C})$, and solve it.

Answer: The Hamiltonian system is given by

$$\frac{dq}{dt} = p, \quad \frac{dp}{dt} = 6q^2 + a. \tag{A.4}$$

By solving the algebraic equation $H = b$ ($b \in \mathbb{C}$) for the unknown variable p, we obtain $p = \sqrt{4q^3 + 2aq + 2b}$. Hence the differential equation to be satisfied by q is given by

$$\frac{dq}{dt} = \sqrt{4q^3 + 2aq + 2b}. \tag{A.5}$$

Thus, $q = q(t)$ is determined as the inverse function of the elliptic integral

$$t = F(q) = \int_{q_0}^{q} \frac{du}{\sqrt{4u^3 + 2au + 2b}}. \tag{A.6}$$

This example is obviously equivalent to the nonlinear equation

$$y'' = 6y^2 + a \qquad (' = \frac{d}{dt}) \tag{A.7}$$

of second order for $y = q$. Recall the differential equation of the Weierstrass function $\wp(t) = \wp(t; \omega_1, \omega_2)$:

$$(\wp'(t))^2 = 4\,\wp(t)^3 - g_2\,\wp(t) - g_3. \tag{A.8}$$

By differentiating this, we obtain

$$\wp''(t) = 6\wp(t)^2 - \frac{g_2}{2}. \tag{A.9}$$

By reparametrizing the constants appropriately, we see that the general solution of the equation (A.7) is given by

$$y(t) = \wp(t - c; \omega_1, \omega_2). \tag{A.10}$$

Hamiltonian systems which arise in the context of Painlevé equations mostly depend explicitly on t, and such systems cannot be integrated by the method described above. In fact, the first Painlevé equation

$$P_{\mathrm{I}}: \qquad y'' = 6y^2 + t \tag{A.11}$$

is the one obtained by replacing the parameter a in (A.7) by the independent variable t. This equation can also be written as a Hamiltonian system

$$\frac{dq}{dt} = p, \quad \frac{dp}{dt} = 6q^2 + t \tag{A.12}$$

with Hamiltonian $H = \frac{1}{2}p^2 - 2q^3 - tq$. It is known, however, that the first Painlevé equation is a transcendental equation that "cannot be integrated".

Generalizing the situation, a Hamiltonian system *of n degrees of freedom* can be defined as a system of differential equations in the form

$$\frac{dq_i}{dt} = \frac{\partial H}{\partial p_i}, \quad \frac{dp_i}{dt} = -\frac{\partial H}{\partial q_i} \qquad (i = 1, \ldots, n) \tag{A.13}$$

for $2n$ dependent variables $(q_1, \ldots, q_n, p_1, \ldots, p_n)$, where the Hamiltonian $H = H(q_1, \ldots, q_n, p_1, \ldots, p_n; t)$ is a function in the variables $(q_1, \ldots, q_n, p_1, \ldots, p_n)$ and t. When we speak about a Hamiltonian system with many time variables $t_1, \ldots, t_m$, we use Hamiltonians $H_1, \ldots, H_m$ in equal number as the time variables, and consider simultaneously the Hamiltonian systems (A.13) for all the time variables t_j with the corresponding Hamiltonians H_j.

A.2. Poisson structure and canonical transformations [Section 1.1]

When φ and ψ are functions of (q,p,t), we define the Poisson bracket $\{\varphi,\psi\}$ by

$$\{\varphi,\psi\} = \frac{\partial\varphi}{\partial p}\frac{\partial\psi}{\partial q} - \frac{\partial\varphi}{\partial q}\frac{\partial\psi}{\partial p}. \tag{A.14}$$

If the variables q,p are subject to a Hamiltonian system with Hamiltonian $H = H(q,p;t)$, then we have

$$\varphi' = \{H,\varphi\} + \frac{\partial\varphi}{\partial t} \qquad ({}' = \frac{d}{dt}) \tag{A.15}$$

for any function $\varphi = \varphi(q,p;t)$ of (q,p,t). Let us consider a coordinate system (λ,μ,s) for the (q,p,t)-space such that $\{\mu,\lambda\} = 1$ and $s = t$. If we transform the Hamiltonian system by such a canonical transformation, at least locally the transformed equation can be expressed again as a Hamiltonian system. We now explain how this is possible. In view of

$$\lambda' = \{H,\lambda\} + \frac{\partial\lambda}{\partial t}, \quad \mu' = \{H,\mu\} + \frac{\partial\mu}{\partial t}, \tag{A.16}$$

we try to find a function f such that

$$\{f,\lambda\} = \frac{\partial\lambda}{\partial t}, \quad \{f,\mu\} = \frac{\partial\mu}{\partial t}. \tag{A.17}$$

With such a function f, setting $K = H + f$ we obtain a Hamiltonian system

$$\lambda' = \{K,\lambda\} = \frac{\partial K}{\partial\mu}, \quad \mu' = \{K,\mu\} = -\frac{\partial K}{\partial\lambda} \tag{A.18}$$

with Hamiltonian K. We first rewrite (A.17) in the form

$$\begin{bmatrix} \dfrac{\partial f}{\partial\lambda} \\ \dfrac{\partial f}{\partial\mu} \end{bmatrix} = \begin{bmatrix} -\dfrac{\partial\mu}{\partial t} \\ \dfrac{\partial\lambda}{\partial t} \end{bmatrix}, \tag{A.19}$$

and then multiply this equality by the Jacobian determinant to get

$$\begin{bmatrix} \dfrac{\partial f}{\partial q} \\ \dfrac{\partial f}{\partial p} \end{bmatrix} = \begin{bmatrix} \dfrac{\partial\lambda}{\partial q} & \dfrac{\partial\mu}{\partial q} \\ \dfrac{\partial\lambda}{\partial p} & \dfrac{\partial\mu}{\partial p} \end{bmatrix} \begin{bmatrix} -\dfrac{\partial\mu}{\partial t} \\ \dfrac{\partial\lambda}{\partial t} \end{bmatrix}. \tag{A.20}$$

Denoting the right-hand side by $[u,v]^{\mathrm{t}}$, we need to prove $\partial u/\partial p = \partial v/\partial q$ in order to guarantee the local existence of f. This is shown by the computation

$$\frac{\partial v}{\partial q} - \frac{\partial u}{\partial p} = \frac{\partial}{\partial t}\left(\frac{\partial\mu}{\partial p}\frac{\partial\lambda}{\partial q} - \frac{\partial\mu}{\partial q}\frac{\partial\lambda}{\partial p}\right) = \frac{\partial}{\partial t}\{\mu,\lambda\} = 0. \tag{A.21}$$

We remark that the Poisson bracket has the following fundamental properties:

(1) The bracket $\{\,,\,\}$ is bilinear and skew-symmetric:

$$\{f,f\} = 0, \quad \{g,f\} = -\{f,g\}.$$

(2) The bracket $\{\,,\,\}$ satisfies the Leibniz rule in each of the two arguments:

$$\{fg,h\} = \{f,h\}g + f\{g,h\}, \quad \{f,gh\} = \{f,g\}h + g\{f,h\}.$$

(3) The bracket $\{\,,\,\}$ satisfies the Jacobi rule:

$$\{f,\{g,h\}\}+\{g,\{h,f\}\}+\{h,\{f,g\}\}=0.$$

A.3. Riccati equations [Section 1.2]

Let us consider an ordinary differential equation

$$y'=a(t)y^2+b(t)y+c(t)\qquad (a(t)\not\equiv 0) \tag{A.22}$$

which requires that the derivative of $y=y(t)$ should be expressed as a quadratic function of y having known functions as coefficients. A first order differential equation of this form is called a *Riccati equation.* We introduce a new dependent variable $u=u(t)$ by setting

$$y=-\frac{1}{a(t)}\frac{d}{dt}\log u=-\frac{1}{a(t)}\frac{u'}{u}. \tag{A.23}$$

Then the differential equation to be satisfied by u is computed as

$$u''-\left(\frac{a'(t)}{a(t)}+b(t)\right)u'+a(t)c(t)u=0. \tag{A.24}$$

Namely, u is determined by a *linear differential equation* of second order. We now take two linearly independent solutions $\varphi_0(t)$, $\varphi_1(t)$ of (A.24), and set

$$u(t)=c_0\,\varphi_0(t)+c_1\,\varphi_1(t)\quad (c_0,c_1\in\mathbb{C}). \tag{A.25}$$

Then we obtain a family of solutions

$$\varphi(t;c_0,c_1)=-\frac{1}{a(t)}\frac{c_0\varphi_0'(t)+c_1\varphi_1'(t)}{c_0\varphi_0(t)+c_1\varphi_1(t)} \tag{A.26}$$

of the Riccati equation (A.23). Since the function $\varphi(t;c_0,c_1)$ is determined by the ratio of c_0 and c_1, this family of solutions is parametrized by the one-dimensional projective space $\mathbb{P}^1$ with homogeneous coordinates $[c_0:c_1]$.

This argument can be summarized as follows. By the transformation (A.23), the Riccati equation (A.22) is *linearized* into the linear differential equation (A.24) of second order. As a result, it turns out that the Riccati equation has a one-dimensional family of solutions parameterized by $\mathbb{P}^1$. It is reasonable to consider that all the solutions of the Riccati equation are obtained in this way.

A.4. Computation of Bäcklund transformations [Sections 1.5 and 2.3]

There are two possible ways in dealing with the Bäcklund transformations. One approach is to regard them as transformations of dependent variables. (In Section 1, we mainly followed this approach.) The other is to regard them as the transformations of solutions. These two approaches are in fact *dual* to each other, and should be distinguished clearly in order to avoid confusion. We explain this point below in some detail.

The same situation arises already on an elementary level. For instance, let us consider parallel translations in the xy plane. The parallel translation by the vector (a,b) transfers a point (x,y) to the point $(x+a,y+b)$. On the other hand, the parabola $y=x^2$ is transferred to the parabola with defining equation $y-b=(x-a)^2$. Namely, on the level of coordinate functions, this parallel translation corresponds to the replacement $x\to x-a$, $y\to y-b$. We first give a general remark of this sort.

In general, let us consider the situation when a group G acts on some space X from the left. Then the unit element $1 \in G$ acts on X as the identity transformation $(1.x = x)$, and for any $g_1, g_2 \in G$ we have

$$(g_1 g_2).\boldsymbol{x} = g_1.(g_2.\boldsymbol{x}) \qquad (\boldsymbol{x} \in X). \tag{A.27}$$

(A mapping $G \times X \to X$, denoted by $(g, x) \mapsto g.x$, is called a *left action* of G on X if it has the property mentioned above.) In this setting, for an element $g \in G$ and a function φ on X we define a function $L_g.\varphi$ on X by

$$(L_g.\varphi)(\boldsymbol{x}) = \varphi(g^{-1}.\boldsymbol{x}) \qquad (\boldsymbol{x} \in X). \tag{A.28}$$

Then we have $L_1.\varphi = \varphi$ and

$$L_{g_1 g_2}.\varphi = L_{g_1}.(L_{g_2}.\varphi) \tag{A.29}$$

for any $g_1, g_2 \in G$. In this way we obtain a left action $(g, \varphi) \mapsto L_g.\varphi$ of G on the set of all functions on X. Note that the composition of L_{g_1} and L_{g_2} is made in the same order as the product of g_1 and g_2. When we work mainly with functions rather than points of a space, we also use the abbreviation $g.\varphi = L_g.\varphi$, just omitting L, in order to simplify the notation. Under this convention, the action of $g \in G$ on a function φ is given by

$$(g.\varphi)(\boldsymbol{x}) = \varphi(g^{-1}.\boldsymbol{x}) \qquad (\boldsymbol{x} \in X). \tag{A.30}$$

Notice that if one replaces g^{-1} by g in (A.28), then L_{g_1} and L_{g_2} should be composed in the opposite order in (A.29). Since the points of a space and the functions on it play complementary roles to each other, it would be natural to use g^{-1} somewhere as long as one wants to make the group G act from the left both on the points and on the functions. If you do not want to use g^{-1} in defining the action on functions, you would need two types of actions, one from the left and the other from the right. Although it is a matter of taste which convention you take, it would cause serious confusion unless you have a clear understanding about what is happening here. In the following, we will follow the convention (A.30), so that the group G should act from the left both on points and on functions. (This convention is more or less standard in the literature, although it is not always the best choice.)

In differential equations, the same situation arises as the relation between the transformation of variables and that of solutions. In this case, the individual solutions correspond to the "points", and the variables appearing in the equation to the "functions". As an example, let us consider the Bäcklund transformations for the second Painlevé equation. Regarding the parameter as a dependent variable, we denote by X the space of all solutions $\boldsymbol{x} = (q_0, p_0; b_0)$ of H_{II}. Then the variables q, p, b are the "functions" on X that extract from each solution $\boldsymbol{x} = (q_0, p_0; b_0)$ the components q_0, p_0, b_0, respectively:

$$q(\boldsymbol{x}) = q_0, \quad p(\boldsymbol{x}) = p_0, \quad b(\boldsymbol{x}) = b_0. \tag{A.31}$$

We now define two transformations of solutions $s, r : X \to X$ by setting

$$\begin{aligned} s.\boldsymbol{x} &= s.(q_0, p_0; b_0) = (q_0 + \frac{b_0}{p_0}, p_0; -b_0), \\ r.\boldsymbol{x} &= r.(q_0, p_0; b_0) = (-q_0, -p_0 + 2q_0^2 + t; 1 - b_0) \end{aligned} \tag{A.32}$$

for each $\boldsymbol{x} = (q_0, p_0; b_0) \in X$, and compute the composition $w = sr$. The point $w.\boldsymbol{x}$ is computed as follows:

$$w.\boldsymbol{x} = s.r(q_0, p_0, b_0) = s.(-q_0, -p_0 + 2q_0^2 + t; 1 - b_0) = \left(-q_0 + \frac{-b_0}{-p_0 + 2q_0^2 + t}, -p_0 + 2q_0^2 + t; -1 + b_0\right). \tag{A.33}$$

Please compare this result with the computation of the composition of Bäcklund transformations in Section 1.4. For comparison, let us compute $w^{-1}.\boldsymbol{x}$. Since the actions of s^2, r^2 on the solution space both reduce to the identity transformation, we have $w^{-1} = rs$. Hence

$$w^{-1}.\boldsymbol{x} = r.s(q_0, p_0, b_0) = r.\left(q_0 + \frac{b_0}{p_0}, p_0; -b_0\right) = \left(-q_0 - \frac{b_0}{p_0}, -p_0 + 2(q_0 + \frac{b_0}{p_0})^2 + t; 1 + b_0\right). \tag{A.34}$$

These transformations of solutions are transferred to those of the dependent variables as follows. Since we have

$$(s.q)(\boldsymbol{x}) = q(s^{-1}.\boldsymbol{x}) = q\left(\left(q_0 + \frac{b_0}{p_0}, p_0; -b_0\right)\right) = q_0 + \frac{b_0}{p_0} = q(\boldsymbol{x}) + \frac{b(\boldsymbol{x})}{p(\boldsymbol{x})} = \left(q + \frac{b}{p}\right)(\boldsymbol{x}), \tag{A.35}$$

we obtain $s.q = q + b/p$, and similarly $s.p = p$, $s.b = -b$. It is because $s^{-1} = s$ that the same formulas have appeared in transformations both of the solutions and of the dependent variables. The difference between the two actions can be seen clearly if we compute the composition $w = sr$. Since $w^{-1} = rs$, we have

$$(w.q)(\boldsymbol{x}) = q(w^{-1}.\boldsymbol{x}) = q(r.s.\boldsymbol{x}) = q\left(\left(-q_0 - \frac{b_0}{p_0}, -p_0 + 2\left(q_0 + \frac{b_0}{p_0}\right)^2 + t; 1 + b_0\right)\right) = -q_0 - \frac{b_0}{p_0} = \left(-q - \frac{b}{p}\right)(\boldsymbol{x}). \tag{A.36}$$

Hence we obtain $w.q = -q - b/p$, and similarly $w.p = -p + 2(q + b/p)^2 + t$, $w.b = 1 + b$. Please check that this computation gives rise to the same result as the computation in Section 1.4. Here we obtained the formulas for $w.q$, $w.p$, $w.b$ from the computation of $w^{-1}.\boldsymbol{x}$. We also remark that the computation of $w.\boldsymbol{x}$ is equivalent to computing the three formulas for $w^{-1}.q$, $w^{-1}.p$, $w^{-1}.b$ simultaneously.

The points of this discussion can be summarized as follows.

(1) The action $w.\varphi$ of w on a function $\varphi = \varphi(q, p, b)$ of the dependent variables q, p, b corresponds to the transformation $w^{-1}.\boldsymbol{x}$ of the solutions $\boldsymbol{x} = (q_0, p_0; b_0)$. If we express $w^{-1}.\boldsymbol{x}$ as $w^{-1}.\boldsymbol{x} = (q_1, p_1; b_1)$, then the three components q_1, p_1, b_1 of $w^{-1}.\boldsymbol{x}$ coincide with the functions obtained from $w.q$, $w.p$, $w.b$ (functions of q, p, b) by substituting the special solution $q = q_0$, $p = p_0$, $b = b_0$. When w is decomposed into the product of several transformations $w = w_1 \cdots w_m$, the computation of $w^{-1}.\boldsymbol{x}$ on the level

of solutions should be carried out by arranging the inverses of w_i in the opposite order: $w.\varphi = w_1.\cdots.w_m.\varphi$, $w^{-1}.\boldsymbol{x} = w_m^{-1}.\cdots.w_1^{-1}.\boldsymbol{x}$.

(2) The transformations of dependent variables can be computed just by formal replacement of variables. For instance, for computing $w.q$ there is no need to keep track of $w.p$ and $w.b$. Once we have completed the computation of $w.q$ and have obtained a formula for it as a function of q, p, b, that computation is applicable to any solution. If we want to know q_1 in the transformation $w^{-1}.\boldsymbol{x} = (q_1, p_1, b_1)$ of a particular solution $\boldsymbol{x} = (q_0, p_0; b_0)$, however, it is only after we have completed the computation of the general formula for $w.q$ that we can substitute each particular solution.

(3) In the transformation on the level of solutions, $w.\boldsymbol{x}$ can be computed by referring only to a particular solution $\boldsymbol{x} = (q_0, p_0; b_0)$. For the purpose of determining $w.\boldsymbol{x}$ for an individual solution, the computation can be done far more effectively compared with the method of (2). However, one needs to keep track of all the three data q, p, b as a vector all the time during the computation, and also the computation itself cannot be applied to other solutions.

It should be noted that each of the two methods has advantages and disadvantages. In practice, the phrase "Bäcklund transformation" is used more often for referring to the transformation on the level of solutions. In this book, however, I decided to put more emphasis on the role of Bäcklund transformations as transformations of dependent variables, in view of its theoretical significance.

A.5. Classical solutions and invariant divisors [Sections 1.6 and 2.2]

According to H. Umemura[3], we define a *classical function* as

> a function obtained by a successive application of the following procedures starting from the rational functions: the arithmetic operations (addition, subtraction, multiplication and division) on known functions, differentiation of known functions, solving algebraic equations whose coefficients are known functions, solving linear ordinary differential equations whose coefficients are known functions, and substituting known functions into abelian functions.

Umemura proposed a general scheme for determining classical solutions (solutions expressible in terms of classical functions) for Painlevé equations. The main point of his scheme was that

> the classification of transcendental classical solutions (classical solutions that are not algebraic functions) can be reduced to the classification of the *invariant divisors*.

(For algebraic solutions, one needs to investigate separately.)

Let us consider a Hamiltonian system

$$\frac{dq}{dt} = \frac{\partial H}{\partial p}, \quad \frac{dp}{dt} = -\frac{\partial H}{\partial q} \tag{A.37}$$

[3] H. Umemura: Painlevé equations and classical functions, Sugaku Expositions **11** (1998), 77–100.

with polynomial Hamiltonian $H = H(q, p; t) \in \mathbb{C}(t)[q, p]$. Let (K, δ) be a differential extension of the field of rational functions $\mathbb{C}(t)$ and define the vector field X acting on the polynomial ring $K[q, p]$ by

$$X = \frac{\partial H}{\partial p}\frac{\partial}{\partial q} - \frac{\partial H}{\partial q}\frac{\partial}{\partial p} + \delta. \tag{A.38}$$

Then a nonzero polynomial $F \in K[q, p]$ is called an *invariant divisor* of the Hamiltonian (A.37) over K if there exists a polynomial $G \in K[q, p]$ such that $X(F) = GF$. Roughly speaking, Umemura's theory implies the following:

> If no nontrivial invariant divisor appears in $K[q, p]$ for any differential extension K of $\mathbb{C}(t)$, then the Hamiltonian system (A.37) has no transcendental classical solution.

In the case of the first Painlevé equation, it is known that for any K there is no nontrivial invariant divisor in $K[q, p]$ (by Kolchin and Nishioka). Combining this with the fact that P_{I} has no algebraic solution, we see that all the solutions of P_{I} are non-classical. The second Painlevé equation has in fact non-trivial invariant divisors. By using Bäcklund transformations, one can reduce the discussion to the case where $0 \leq \mathrm{Re}\, b < 1$. When the parameter b belongs to this region, one can show that non-trivial invariant divisors appear only in the case $b = 0$, and that all the invariant divisors for $b = 0$ are expressed as $c\,p^n$ ($c \in K^*$, $n = 0, 1, 2, \ldots$). From this, it follows that H_{II} has transcendental classical solutions only for integer values of b, and that all the transcendental classical solutions of H_{II} are obtained from the Riccati solution at $b = 0$ by Bäcklund transformations. On the other hand, it is known that the algebraic solutions of H_{II} are all rational, and that all the rational solutions are obtained from the one at $b = 1/2$ by Bäcklund transformations.

The classification of invariant divisors has now been completed for all the six Painlevé equations (Okamoto-Noumi, Murata, Umemura-Watanabe). Also, for the five Painlevé equations except P_{VI} the classification of algebraic solutions has been completed; in that sense, all the classical solutions are determined already for the five equations. For P_{VI} however, the whole picture of algebraic solutions has not been clarified yet.

A.6. Semidirect product of groups [Section 2.3]

Let G be a group. A bijection $\varphi : G \to G$ satisfying $\varphi(1) = 1$ and $\varphi(g_1 g_2) = \varphi(g_1)\varphi(g_2)$ is called an *automorphism* of G. We denote by $\mathrm{Aut}(G)$ the group of all automorphisms of G, and call it the automorphism group of G. For each $g \in G$, the mapping $\mathrm{Ad}(g) : G \to G$ defined by

$$\mathrm{Ad}(g)(x) = g\,x\,g^{-1} \quad (x \in G) \tag{A.39}$$

is an automorphism of G. An automorphism of this form is called an *inner automorphism* of G. From this construction, we obtain a homomorphism of groups

$$\mathrm{Ad}:\ G \to \mathrm{Aut}(G):\ \ g \mapsto \mathrm{Ad}(g). \tag{A.40}$$

(Note that the kernel of Ad coincides with the center of G.)

Suppose now that G is contained in a bigger group $\widetilde{G}$ as a *normal* subgroup. Suppose in addition that there exists a subgroup $\Omega \subset \widetilde{G}$ such that

$$\widetilde{G} = G\,\Omega, \quad G \cap \Omega = \{1\}. \tag{A.41}$$

If this is the case, $\widetilde{G}$ is said to be a *semidirect product* of G and Ω, and is denoted as $\widetilde{G} = G \rtimes \Omega$. This condition implies that $\widetilde{G}/G \simeq \Omega$ and that any element $\widetilde{g} \in \widetilde{G}$ can be expressed uniquely in the form $\widetilde{g} = g\omega$ $(g \in G, \omega \in \Omega)$. Then the product of two elements $\widetilde{g_1} = g_1\omega_1$, $\widetilde{g_2} = g_2\omega_2$ of G is computed as

$$\widetilde{g_1}\widetilde{g_2} = g_1\omega_1 g_2\omega_2 = g_1(\omega_1 g_2 \omega_1^{-1})\omega_1\omega_2. \tag{A.42}$$

Since G is a normal subgroup, we have

$$\mathrm{Ad}(\omega_1)(g_2) = \omega_1 g_2 \omega_1^{-1} \in G. \tag{A.43}$$

Hence the product $\widetilde{g} = \widetilde{g_1}\widetilde{g_2}$ is decomposed into $\widetilde{g} = g\omega$ $(g \in G, \omega \in \Omega)$, where

$$g = g_1 \mathrm{Ad}(\omega_1)(g_2), \quad \omega = \omega_1\omega_2. \tag{A.44}$$

Conversely, suppose that the two groups G and Ω are given, and that Ω acts on G as a group of automorphisms. Then one can construct a bigger group $\widetilde{G}$ that has the property mentioned above. Let

$$\rho : \Omega \to \mathrm{Aut}(G) \tag{A.45}$$

be the homomorphism of groups that specifies the action of ω on G as a group of automorphisms. Then one can construct a new group by introducing the following product on the direct product $G \times \Omega$ of sets: For each (g_1, ω_1), $(g_2, \omega_2) \in G \times \Omega$

$$(g_1, \omega_1).(g_2, \omega_2) = (g_1\, \rho(\omega_1)(g_2), \omega_1\omega_2). \tag{A.46}$$

This group is denoted by $G \rtimes_\rho \Omega$, and called the *semidirect product* of G with Ω acting on G. When Ω is given as a subgroup of $\mathrm{Aut}(G)$, or when it is clear from the context which homomorphism ρ is relevant, the semidirect product $G \rtimes_\rho \Omega$ is simply denoted by $G \rtimes \Omega$.

Take the subsets $\hat{G} = G \times \{1\}$, $\hat{\Omega} = \{1\} \times \Omega$ of $G \rtimes_\rho \Omega$. Then $\hat{G}$ and $\hat{\Omega}$ are subgroups of $G \rtimes_\rho \Omega$, and $\hat{G}$ is normal. Since $G \rtimes_\rho \Omega = \hat{G}\hat{\Omega}$ and $\hat{G} \cap \hat{\Omega} = \{1\}$, the group $\widetilde{G} = G \rtimes_\rho \Omega$ is the semidirect product of $\hat{G}$ and $\hat{\Omega}$ in the sense mentioned earlier.

A.7. Cartan matrices and Dynkin diagrams [Section 2.3]

A square matrix $A = (a_{ij})_{i,j \in I}$ with integer entries is called a *generalized Cartan matrix* (*GCM* for short) or a *Kac-Moody matrix* if it satisfies the following three conditions:

(1) $a_{ii} = 2 \quad (i \in I)$.
(2) $a_{ij} \leq 0 \quad (i, j \in I,\ i \neq j)$.
(3) $a_{ij} = 0 \iff a_{ji} = 0$.

In this book we abbreviate "generalized Cartan matrix" simply as "Cartan matrix".

For such a matrix $A = (a_{ij})_{i,j \in I}$, we define the Dynkin diagram of A by the following procedure. To each index $i \in I$, we assign a node (circle) $\circ$ and put the label i near it if necessary. For each pair of indices (i, j) with $(a_{ij}, a_{ji}) \neq (0, 0)$ we join the two nodes corresponding to i and j by the following rule:

(A.47)

$a_{ij}a_{ji}$	(a_{ij}, a_{ji})	i j
0	$(0,0)$	
1	$(-1,-1)$	
2	$(-2,-1)$	
3	$(-3,-1)$	
4	$(-2,-2)$	
4	$(-4,-1)$	

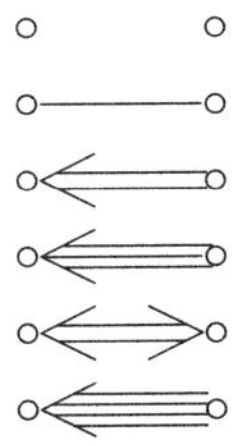

We will not specify here the rule for the cases when $a_{ij}a_{ji} > 4$.

The *Weyl group* $W = W(A)$ associated with the generalized Cartan matrix A is the group defined by generators s_i $(i \in I)$ and fundamental relations

$$s_i^2 = 1 \quad (i \in I); \quad (s_i s_j)^{m_{ij}} = 1 \quad (i, j \in I, \ i \neq j).$$

Here the exponents m_{ij} are defined as follows according to the value of the product $a_{ij}a_{ji}$:

$a_{ij}a_{ji}$	0	1	2	3	≥ 4
m_{ij}	2	3	4	6	∞

When $m_{ij} = \infty$, we do not impose any condition on $s_i s_j$ for such a pair (i, j). The relations of the form $(s_i s_j)^{m_{ij}} = 1$ are often called *Coxeter relations*. Note that, in the case of a Weyl group, the exponents m_{ij} are specified by the generalized Cartan matrix A. Since $s_i^2 = 1$ $(i \in I)$, the relation $(s_i s_j)^{m_{ij}} = 1$ is equivalently rewritten as follows:

(A.48)
$$\begin{aligned}
a_{ij}a_{ji} = 0 \quad &\Rightarrow \quad s_i s_j = s_j s_i, \\
a_{ij}a_{ji} = 1 \quad &\Rightarrow \quad s_i s_j s_i = s_j s_i s_j, \\
a_{ij}a_{ji} = 2 \quad &\Rightarrow \quad s_i s_j s_i s_j = s_j s_i s_j s_i, \\
a_{ij}a_{ji} = 3 \quad &\Rightarrow \quad s_i s_j s_i s_j s_i s_j = s_j s_i s_j s_i s_j s_i.
\end{aligned}$$

These relations are called the *braid relations*. Although the Coxeter relations and the braid relations are equivalent to each other under the condition $s_i^2 = 1$ $(i \in I)$, the braid relations are sometimes more useful when we need to consider more general algebraic structures.

Let Ω be a group of permutations of the indices (a subgroup of $\mathrm{Aut}(I)$), and suppose that

(A.49) $$a_{\omega(i)\omega(j)} = a_{ij} \quad (i, j \in I)$$

for any $\omega \in \Omega$. Such an ω is called simply an *automorphism of the Dynkin diagram* for the obvious reason. If we set

(A.50) $$\rho(\omega)(s_i) = s_{\omega(i)} \quad (i \in I)$$

for each $\omega \in \Omega$, the group Ω acts on the Weyl group $W = W(A)$ through the homomorphism $\rho : \Omega \to \mathrm{Aut}(W)$. The semidirect product $\widetilde{W} = W \rtimes_\rho \Omega$ is called an *extended* Weyl group associated with the group of diagram automorphisms Ω. If we identify $\omega \in \Omega$ with $(1, \omega) \in \widetilde{W}$, the action of $\rho(\omega)$ gives rise to the relations

(A.51) $$\omega s_i = s_{\omega(i)} \omega \quad (i \in I)$$

in $\widetilde{W}$.

The generalized Cartan matrices are classified into three types: *finite type*, *affine type* and *indefinite type*. When the generalized Cartan matrix A is of finite type, the corresponding Weyl group W is a finite group. When A is of affine type, W is an infinite group called an *affine Weyl group*. In the affine case, W is decomposed into the semidirect product of a lattice (a free abelian group of finite rank) and a finite Weyl group acting on it[4].

For a given generalized Cartan matrix $A = (a_{ij})_{i,j\in I}$, consider the two free $\mathbb{Z}$-modules

(A.52) $$Q = \bigoplus_{i\in I} \mathbb{Z}\,\alpha_i, \quad Q^\vee = \bigoplus_{i\in I} \mathbb{Z}\,h_i$$

and the pairing

(A.53) $$\langle\ ,\ \rangle:\ Q^\vee \times Q \to \mathbb{Z}; \quad \langle\, h_i, \alpha_j \,\rangle = a_{ij} \quad (i, j \in I)$$

between $Q^\vee$ and Q. When we regard A as the table of a pairing (bilinear mapping) in this way, Q and $Q^\vee$ are called the *root lattice* and the *coroot lattice*, respectively; the elements α_i and h_i $(i \in I)$ are called the *simple roots* and the *simple coroots*, respectively. The Weyl group $W = W(A)$ acts canonically on the root and the coroot lattices. The action of s_i on Q, $Q^\vee$ is defined by

(A.54) $$s_i(\beta) = \beta - \langle\, h_i, \beta \,\rangle \alpha_i \ \ (\beta \in Q); \quad s_i(h) = h - \langle\, h, \alpha_i \,\rangle h_i \ \ (h \in Q^\vee).$$

These s_i give a realization of the Weyl group $W = W(A)$ as a group of linear transformations on Q and $Q^\vee$, respectively; in particular, the linear transformations s_i satisfy the defining relations of $W(A)$. Furthermore, the bilinear mapping $\langle\ ,\ \rangle$ is W-invariant in the sense that

(A.55) $$\langle\, w.h, w.\beta \,\rangle = \langle\, h, \beta \,\rangle \qquad (h \in Q^\vee, \beta \in Q).$$

Consider a pair (h, α) of $\alpha \in Q$ and $h \in Q^\vee$ such that $\langle\, h, \alpha \,\rangle = 2$. For such (h, α), we define the *reflection* $s_{(h,\alpha)}$ *with respect to the pair* (h, α) by

(A.56) $$s_{(h,\alpha)}(\beta) = \beta - \langle\, h, \beta \,\rangle \alpha \quad (\beta \in Q).$$

The reflections s_i with respect to (h_i, α_i), defined above, are called the *simple reflections*. In general, a element $\alpha \in Q$ is called a *real root* if it belongs to the W-orbit of some simple root, namely, if it is expressed as $\alpha = w.\alpha_i$ for some $i \in I$ and $w \in W$. For each real root $\alpha = w.\alpha_i$, the element $s_\alpha = w s_i w^{-1} \in W$ obtained from s_i by conjugation is the reflection with respect to the pair (h, α), where $h = w.h_i$.

A.8. Demazure operators [Section 2.5]

Let $f = f(x, y)$ be a polynomial in x, y. The following is a way to produce a symmetric polynomial with respect to (x, y) from f. Define $\Delta(f)$ by

(A.57) $$\Delta(f)(x, y) = \frac{f(y, x) - f(x, y)}{y - x}.$$

---

[4] For various aspects of generalized Cartan matrices, including the relation with Kac-Moody Lie algebras, see the textbooks by V.G. Kac and by M. Wakimoto [V.G. Kac, *Infinite Dimensional Lie Algebras*, Third Edition, Cambridge University Press, 1990; M. Wakimoto: *Infinite-Dimensional Lie Algebras*, Translations of Mathematical Monographs 195, American Mathematical Society, 2001].

Since the denominator is alternating with respect to (x, y), it is divisible by $y - x$, and the resulting quotient is a symmetric polynomial. For example, we have

$$\Delta(x^n) = \frac{y^n - x^n}{y - x} = x^{n-1} + x^{n-2}y + \cdots + y^{n-1}. \tag{A.58}$$

The polynomial $\Delta(f)$ is called the *divided difference* of f. (This construction probably goes back to Newton.) Since $\Delta(f) = 0$ whenever f is symmetric, we see that $\Delta^2(f) = 0$ for any polynomial f. Also, for the product of functions, $\Delta(fg)$ can be computed by the twisted Leibniz rule. In fact, we have

$$\begin{aligned} &f(y,x)g(y,x) - f(x,y)g(x,y) \\ &= (f(y,x) - f(x,y))g(x,y) + f(y,x)(g(y,x) - g(x,y)), \end{aligned} \tag{A.59}$$

and hence

$$\Delta(fg) = \Delta(f)g + s(f)\Delta(g), \tag{A.60}$$

where $s(f)$ stands for the polynomial defined by $s(f)(x,y) = f(y,x)$. Similar operations can be formulated as Demazure operators in terms of the root system and the Weyl group.

Consider an operator s with $s^2 = 1$, and take a variable α such that $s(\alpha) = -\alpha$. (In the context above, $\alpha = y - x$.) Then we define the *Demazure operator* Δ by setting

$$\Delta(f) = \frac{1}{\alpha}(s(f) - f). \tag{A.61}$$

This operator has the following two properties:

(1) If f is s-invariant (i.e. $s(f) = f$), then $\Delta(f) = 0$.

(2) For the product of functions,

$$\Delta(fg) = \Delta(f)g + s(f)\Delta(g) = \Delta(f)s(g) + f\Delta(g). \tag{A.62}$$

In particular, one has $\Delta(fg) = f\Delta(g)$ when f is s-invariant.

Let $A = (a_{ij})_{i,j \in I}$ be a generalized Cartan matrix, and consider the situation where the Weyl group $W = W(A)$ associated with A acts on a commutative algebra $\mathcal{R}$ as a group of automorphisms. We assume in addition that $\mathcal{R}$ contains invertible elements α_i $(i \in I)$ such that

$$s_i(\alpha_j) = \alpha_j - \alpha_i\langle h_i, \alpha_j \rangle = \alpha_j - \alpha_i a_{ij}. \tag{A.63}$$

Then for each $i \in I$, we define the Demazure operator Δ_i by

$$\Delta_i = \frac{1}{\alpha_i}(s_i - 1). \tag{A.64}$$

Then we have

(1) If $f \in \mathcal{R}$ and $s_i(f) = f$, then $\Delta_i(f) = 0$.

(2) For any $f, g \in \mathcal{R}$,

$$\Delta_i(fg) = \Delta_i(f)g + s_i(f)\Delta_i(g) = \Delta_i(f)s_i(g) + f\Delta_i(g). \tag{A.65}$$

In particular, $\Delta_i(fg) = f\Delta_i(g)$ whenever f is s_i-invariant.

(3) For each $i \in I$, $\Delta_i^2 = 0$. When $i.j \in I$, $i \neq j$,

$$
\text{(A.66)} \qquad
\begin{aligned}
a_{ij}a_{ji} = 0 &\Rightarrow \Delta_i\Delta_j = \Delta_j\Delta_i,\\
a_{ij}a_{ji} = 1 &\Rightarrow \Delta_i\Delta_j\Delta_i = \Delta_j\Delta_i\Delta_j,\\
a_{ij}a_{ji} = 2 &\Rightarrow \Delta_i\Delta_j\Delta_i\Delta_j = \Delta_j\Delta_i\Delta_j\Delta_i,\\
a_{ij}a_{ji} = 3 &\Rightarrow \Delta_i\Delta_j\Delta_i\Delta_j\Delta_i\Delta_j = \Delta_j\Delta_i\Delta_j\Delta_i\Delta_j\Delta_i.
\end{aligned}
$$

In the setting of Section 2.5, we explain below how one can use Demazure operators to prove that the action of each s_i gives rise to a Bäcklund transformation for the symmetric form of P_{IV}. By using the Cartan matrix $A = (a_{ij})_{i,j=0}^2$ and the matrix $U = (u_{ij})_{i,j=0}^2$, the action of the Demazure operator Δ_i is expressed as follows:

$$
\text{(A.67)} \qquad \Delta_i(\alpha_j) = -a_{ij}, \qquad \Delta_i(f_j) = \frac{1}{f_i}u_{ij}.
$$

Let us write the equations of the symmetric form in the form

$$
\text{(A.68)} \qquad f_j' = F_j(\alpha_0, \alpha_1, \alpha_2; f_0, f_1, f_2) \qquad (j = 0, 1, 2).
$$

Then we have $s_i(f_j') = s_i(F_j)$. On the other hand, we have

$$
\text{(A.69)} \qquad s_i(f_j)' = \left(f_j + \frac{\alpha_i}{f_i}u_{ij}\right)' = F_j - \frac{\alpha_i}{f_i^2}F_i u_{ij}.
$$

Hence we have to show that $s_i(F_j) = F_j - \frac{\alpha_i}{f_i^2}F_i u_{ij}$, namely,

$$
\text{(A.70)} \qquad \Delta_i(F_j) = -\frac{F_i}{f_i^2}u_{ij}.
$$

Since $F_1 = f_1(f_2 - f_0) + \alpha_1$, $\Delta_0(F_1)$ is computed as

$$
\text{(A.71)} \qquad
\begin{aligned}
\Delta_0(F_1) &= \Delta_0(f_1)(f_2 - f_0) + s_0(f_1)\Delta_0(f_2 - f_0) + \Delta_0(\alpha_1)\\
&= \frac{1}{f_0}(f_2 - f_0) + \left(f_1 + \frac{\alpha_0}{f_0}\right)\frac{-1}{f_0} + 1\\
&= \frac{1}{f_0}\left(f_2 - f_1 - \frac{\alpha_0}{f_0}\right) = -\frac{1}{f_0^2}(f_0(f_1 - f_2) + \alpha_0) = -\frac{F_0}{f_0^2}.
\end{aligned}
$$

The other $\Delta_i(F_j)$ can be computed similarly.

A.9. Hirota's bilinear operators [Section 3.2]

We consider a differential operator $P = P(t; \partial_t)$ with respect to a variable t:

$$
\text{(A.72)} \qquad P(t; \partial_t) = \sum_{k=0}^{m} a_k(t)\partial_t^k,
$$

where $\partial_t = d/dt$ stands for the derivative with respect to t, and $a_k(t)$ for the multiplication operator. For a given pair (f, g) of functions $f = f(t)$, $g = g(t)$, we denote the function

$$
\text{(A.73)} \qquad \sum_{k=0}^{m} a_k(t) \sum_{i+j=k} (-1)^j \binom{k}{j} \partial_t^i(f)\partial_t^j(g)
$$

symbolically by

$$
\text{(A.74)} \qquad P(t; D_t)\,(f \cdot g), \quad \text{or} \quad P(t; D_t)\, f \cdot g,
$$

and call $P(t; D_t)$ the *Hirota bilinear operator*. For example, we obtain the following Hirota derivatives:

$$\begin{aligned} 1(f \cdot g) &= fg, \quad D_t(f \cdot g) = f'g - fg', \\ D_t^2(f \cdot g) &= f''g - 2f'g' + fg'', \\ D_t^3(f \cdot g) &= f'''g - 3f''g' + 3f'g'' - g'''. \end{aligned} \tag{A.75}$$

In order to compute the formula for $D^n(f \cdot g)$, one has only to apply the Leibniz rule, using the minus sign as many times as we differentiate the function on the right side. When $f = g$ and k is odd, we always have $D_t^k(f \cdot g) = 0$. If k is even, we have for example

$$\begin{aligned} D_t^2(f \cdot f) &= 2\left(f''f - (f')^2\right), \\ D_t^4(f \cdot f) &= 2\left(f''''f - 4f'''f' + 3(f'')^2\right). \end{aligned} \tag{A.76}$$

These formulas can also be written as

$$\frac{D_t(f \cdot g)}{fg} = \left(\log\left(\frac{f}{g}\right)\right)', \quad \frac{1}{2}\frac{D_t^2(f \cdot f)}{f^2} = (\log f)'' \tag{A.77}$$

in terms of the logarithmic derivatives. Note that the Hirota derivative represents a single function. Once we write a formula like $D_t(f \cdot f) = 2(f''f - (f')^2)$, there is no distinction between the left and the right components. It does not make sense to differentiate such a formula any more in the sense of Hirota.

The definition of $P(t; D_t)(f \cdot g)$ mentioned above can be expressed alternatively as

$$\begin{aligned} P(t; D_t)(f \cdot g) &= \left[P(x; \partial_x - \partial_y) f(x) g(y)\right]_{x=y=t} \\ &= \left[P(t + x; \partial_x) f(t + x) g(t - x)\right]_{x=0}. \end{aligned} \tag{A.78}$$

Also the Taylor expansion of the product $f(t+x)g(t-x)$ with respect to x can be expressed as

$$f(t + x)g(t - x) = \sum_{n=0}^{\infty} \frac{x^n}{n!} D_t^n(f(t) \cdot g(t)) \tag{A.79}$$

in terms of the Hirota derivatives. (This formula can be used as the definition of $D_t^n(f \cdot g)$.) The Hirota bilinear operators for many variables can be formulated similarly. One has only to interpret the definition above as referring to the variables $t = (t_1, t_2, \ldots)$ and the partial derivatives $\partial_t = (\partial_{t_1}, \partial_{t_2}, \ldots)$.

A.10. Application of the exterior algebra [Sections 6.1 and 6.4]

Let $V = \mathbb{C}v_1 \oplus \mathbb{C}v_2 \oplus \cdots \oplus \mathbb{C}v_n$ be the n-dimensional vector space with canonical basis $\{v_1, v_2, \ldots, v_n\}$. The *exterior algebra* $\Lambda(V)$ associated with V is the (non-commutative associative) algebra defined by the generators $v_1, \ldots, v_n$ and the following commutation relations: Denoting by $\wedge$ the product in $\Lambda(V)$,

$$v_j \wedge v_j = 0, \quad v_i \wedge v_j + v_j \wedge v_i = 0 \tag{A.80}$$

for each $i, j = 1, \ldots, n$. One can easily verify that similar relations hold for any

$$f = v_1 a_1 + \cdots + v_n a_n, \quad g = v_1 b_1 + \cdots + v_n b_n \tag{A.81}$$

of V, namely,

$$f \wedge f = 0, \quad f \wedge g + g \wedge f = 0 \tag{A.82}$$

in $\Lambda(V)$. By counting the degree of each element of V as 1, we see that $\Lambda(V)$ is decomposed as follows by the degree:

(A.83) $$\Lambda(V) = \bigoplus_{r=0}^{n} \Lambda^r(V), \quad \Lambda^r(V) = \bigoplus_{1 \le j_1 < \ldots < j_r \le n} \mathbb{C} v_{j_1} \wedge \cdots \wedge v_{j_r}.$$

In particular, the dimension of the r-th component $\Lambda^r(V)$ is $\binom{n}{r}$, and that of the whole space $\Lambda(V)$ is 2^n. As in Chapter 6, for each subset $J = \{j_1, \ldots, j_r\}$ we set

(A.84) $$v_J = v_{j_1} \wedge \cdots \wedge v_{j_r}$$

by arranging the elements in increasing order $j_1 < \ldots < j_r$. In this notation, the set of $\binom{n}{r}$ vectors v_J ($|J| = r$) forms a basis of $\Lambda^r(V)$.

Now let $X = (x_{ij})_{ij}$ be an $n \times n$ matrix, and define the linear mapping $\rho_X : V \to V$ by

(A.85) $$\rho_X(v_j) = \sum_{i=1}^{n} v_i \, x_{ij} \quad (j = 1, \ldots, n).$$

Note that the elements $\widetilde{v}_j = \rho_X(v_j)$ again satisfy the relations

(A.86) $$\widetilde{v}_j \wedge \widetilde{v}_j = 0, \quad \widetilde{v}_i \wedge \widetilde{v}_j + \widetilde{v}_j \wedge \widetilde{v}_i = 0,$$

as we remarked before. Hence $\rho_X : V \to V$ is extended uniquely to an automorphism

(A.87) $$\rho_X : \Lambda(V) \to \Lambda(V)$$

of the exterior algebra. (We have denoted the extension of ρ_X by the same symbol.) Since ρ_X preserves the degree, it induces a linear transformation on each $\Lambda^r(V)$. We now look at the representation matrix of the $\rho_X : \Lambda^r(V) \to \Lambda^r(V)$ for each $r = 0, 1, \ldots, n$:

(A.88) $$\begin{aligned} \rho_X(v_J) &= \rho_X(v_{j_1} \wedge \cdots \wedge v_{j_r}) \\ &= \rho_X(v_{j_1}) \wedge \cdots \wedge \rho_X(v_{j_r}) \\ &= \sum_{k_1, \ldots, k_1} v_{k_1} \wedge \cdots \wedge v_{k_r} \, x_{k_1 j_1} \cdots x_{k_r j_r}. \end{aligned}$$

Note that the vector $v_{k_1} \wedge \cdots \wedge v_{k_r}$ is nonzero if and only if there exist a subset $I = \{i_1, \ldots, i_r\}$ ($i_1 < \ldots < i_r$) and a permutation $\sigma \in \mathfrak{S}_r$ such that

(A.89) $$v_{k_1} = v_{i_{\sigma(1)}}, \quad \ldots, \quad v_{k_r} = v_{i_{\sigma(r)}}.$$

Also, I and σ are determined uniquely in this case. Then we have

(A.90) $$\begin{aligned} & v_{k_1} \wedge \cdots \wedge v_{k_r} \, x_{k_1 j_1} \cdots x_{k_r j_r} \\ &= v_{i_{\sigma(1)}} \wedge \cdots \wedge v_{i_{\sigma(r)}} \, x_{i_{\sigma(1)} j_1} \cdots x_{i_{\sigma(r)} j_r} \\ &= v_{i_1} \wedge \cdots \wedge v_{i_r} \, \epsilon(\sigma) x_{i_{\sigma(1)} j_1} \cdots x_{i_{\sigma(r)} j_r}, \end{aligned}$$

since as many minus signs as the number of inversions in σ are picked up in the process of rearranging. Finally, by summing up we obtain

(A.91) $$\begin{aligned} \rho_X(v_J) &= \rho_X(v_{j_1} \wedge \cdots \wedge v_{j_r}) \\ &= \sum_{i_1 < \ldots < i_r} v_{i_1} \wedge \cdots \wedge v_{i_r} \sum_{\sigma \in \mathfrak{S}_r} \epsilon(\sigma) x_{i_{\sigma(1)} j_1} \cdots x_{i_{\sigma(r)} j_r} \\ &= \sum_{i_1 < \ldots < i_r} v_{i_1} \wedge \cdots \wedge v_{i_r} \, \xi^{i_1 \ldots i_r}_{j_1 \ldots j_r} = \sum_{|I| = r} v_I \, \xi^I_J. \end{aligned}$$

Namely, for each $r = 0, 1, \ldots, n$, the representation matrix of $\rho_X : \Lambda^r(V) \to \Lambda^r(V)$ with respect to the basis $\{v_J\}_{|J|=r}$ is given by the $\binom{n}{r} \times \binom{n}{r}$ matrix

$$\Xi_X^r = \left(\xi_J^I\right)_{|I|=r, |J|=r} \tag{A.92}$$

of all $r \times r$ minor determinants. In this framework, we prove below Theorems 6.1 and 6.9 of Chapter 6.

Formula concerning the product of matrices

Let X and Y be two $n \times n$ matrices. Then we have the equality $\rho_{XY} = \rho_X \rho_Y$ of transformations on V. Each side of this equality is extended uniquely to an algebra automorphism of $\Lambda(V)$, and the two sides coincide with each other on V. Hence they give the same linear transformation on $\Lambda(V)$, and hence on $\Lambda^r(V)$ for each $r = 0, 1, \ldots, n$. The equality $\rho_{XY} = \rho_X \rho_Y$ on $\Lambda^r(V)$ implies the equality of the representation matrices

$$\Xi_{XY}^r = \Xi_X^r \Xi_Y^r \quad \text{(product of matrices)}. \tag{A.93}$$

Since $\Xi_X^r = \left(\xi_J^I\right)_{|I|=r, |J|=r}$, this is equivalent to

$$\xi_J^I(XY) = \sum_{|K|=r} \xi_K^I(X)\, \xi_J^K(Y) \quad (|I| = |J| = r), \tag{A.94}$$

which proves Theorem 6.1. (Although Theorem 6.1 is formulated as a statement for rectangular matrices, there is no essential difference.)

Formula concerning the division of the index set

We now take two subsets $J_1, J_2 \subset \{1, \ldots, n\}$, and set $|J_1| = r_1$, $|J_2| = r_2$, and $r = r_1 + r_2$. Note that we have $v_{J_1} \wedge v_{J_2} = 0$ whenever $J_1 \cap J_2 \neq \emptyset$. If $J_1 \cap J_2 = \emptyset$, then by using as many transpositions as the inversion number $\ell(J_1; J_2)$ one can rewrite $v_{J_1} \wedge v_{J_2}$ into $v_{J_1 \cup J_2}$. This means that

$$v_{J_1} \wedge v_{J_2} = \epsilon(J_1; J_2)\, v_{J_1 \cup J_2}. \tag{A.95}$$

In the following we define $\epsilon(J_1; J_2) = 0$ when $J_1 \cap J_2 \neq \emptyset$. By applying ρ_X we compute

$$\begin{aligned} \epsilon(J_1; J_2)\, \rho_X(v_{J_1 \cup J_2}) &= \rho_X(v_{J_1}) \wedge \rho_X(v_{J_2}) \\ &= \left(\sum_{|I_1|=r_1} v_{I_1} \xi_{J_1}^{I_1} \right) \wedge \left(\sum_{|I_2|=r_2} v_{I_2} \xi_{J_2}^{I_2} \right) \\ &= \sum_{|I|=r_1,\ |I_2|=r_2} v_{I_1} \wedge v_{I_2}\, \xi_{J_1}^{I_1} \xi_{J_2}^{I_2} \\ &= \sum_{|I_1|=r_1\ |I_2|=r_2} \epsilon(I_1; I_2)\, v_{I_1 \cup I_2}\, \xi_{J_1}^{I_1} \xi_{J_2}^{I_2} \\ &= \sum_{|I|=r} v_I \sum_{I_1 \cup I_2 = I} \epsilon(I_1; I_2)\, \xi_{J_1}^{I_1} \xi_{J_2}^{I_2}. \end{aligned} \tag{A.96}$$

On the other hand, we have

$$\rho_X(v_{J_1 \cup J_2}) = \sum_{|I|=r} v_I\, \xi_{J_1 \cup J_2}^I \tag{A.97}$$

if $J_1 \cap J_2 = \emptyset$. By comparing the two formulas, for any subset $I \subset \{1, \dots, n\}$ with $|I| = r$ we obtain the equality

$$\epsilon(J_1; J_2)\, \xi^I_{J_1 \cup J_2} = \sum_{I_1 \cup I_2 = I} \epsilon(I_1; I_2)\, \xi^{I_1}_{J_1} \xi^{I_2}_{J_2}, \tag{A.98}$$

which proves Theorem 6.9. (Theorem 6.9 is in fact a transposed version of this equality.)

Bibliography

[1] R. Conte (Editor): *The Painlevé Property — One Century Later*, CRM Series in Mathematical Physics, Springer, 1999.

[2] V.I. Gromak, I. Laine and S. Shimomura: *Painlevé Differential Equations in the Complex Plane*, de Gruyter Studies in Mathematics 28, Walter de Gruyter & Co., 2002.

[3] E.L. Ince: *Ordinary Differential Equations* (1926), Dover Publications, 1956 .

[4] K. Iwasaki, H. Kimura, S. Shimomura and M. Yoshida: *From Gauss to Painlevé — A Modern Theory of Special Functions*, Aspects of Mathematics E16, Vieweg, 1991.

[5] M. Jimbo: *Holonomic Quantum Field Theory*, Developments of Modern Mathematics 4, Iwanami Shoten Publishers, 1998 (in Japanese).

[6] V.G. Kac: *Infinite Dimensional Lie Algebras*, Third Edition, Cambridge University Press, 1990.

[7] T. Kawai and Y. Takei: *Algebraic Analysis of Singular Perturbation*, Developments of Modern Mathematics 2, Iwanami Shoten Publishers, 1998 (in Japanese).

[8] I.G. Macdonald: *Symmetric Functions and Hall Polynomials*, Second edition, Oxford Mathematical Monographs, Oxford University Press, 1995.

[9] T. Miwa, M. Jimbo and E. Date: *Solitons — Differential Equations, Symmetries and Infinite-Dimensional Algebras*, Iwanami Series in Applied Mathematics, Iwanami Shoten Publishers, 1993 (in Japanese); Cambridge Tracts in Mathematics, 135. Cambridge University Press, 2000 (Translation by M. Reid).

[10] M. Noumi and Y. Yamada: Symmetries in Painlevé equations, Sugaku **53** (2001), 62–75 (in Japanese); Translation to appear in Sugaku Expositions.

[11] K. Okamoto: *Introduction to Painlevé Equations*, Sophia Kokyuroku in Mathematics No. 19, 1985 (in Japanese).

[12] H. Umemura: On the irreducibility of Painlevé differential equations, Sugaku **40** (1988), 47–61 (in Japanese); Translation in Sugaku Expositions **2** (1989), 231–252.

[13] H. Umemura: Painlevé equations and classical functions, Sugaku **47** (1995), 341–359 (in Japanese); Translation in Sugaku Expositions **11** (1998), 77–100.

[14] H. Umemura: 100 years of the Painlevé equations, Sugaku **51** (1999), 395–420 (in Japanese).

[15] M. Wakimoto: *Infinite-Dimensional Lie Algebras*, Developments of Modern Mathematics 3, Iwanami Shoten Publishers, 1999 (in Japanese); Translations of Mathematical Monographs 195, American Mathematical Society, 2001 (Translation by K. Iohara).

Index

Titles in This Series

223 **Masatoshi Noumi,** Painlevé equations through symmetry, 2004
222 **G. G. Magaril-Il'yaev and V. M. Tikhomirov,** Convex analysis: Theory and applications, 2003
221 **Katsuei Kenmotsu,** Surfaces with constant mean curvature, 2003
220 **I. M. Gelfand, S. G. Gindikin, and M. I. Graev,** Selected topics in integral geometry, 2003
219 **S. V. Kerov,** Asymptotic representation theory of the symmetric group and its applications to analysis, 2003
218 **Kenji Ueno,** Algebraic geometry 3: Further study of schemes, 2003
217 **Masaki Kashiwara,** D-modules and microlocal calculus, 2003
216 **G. V. Badalyan,** Quasipower series and quasianalytic classes of functions, 2002
215 **Tatsuo Kimura,** Introduction to prehomogeneous vector spaces, 2003
214 **L. Š. Grinblat,** Algebras of sets and combinatorics, 2002
213 **V. N. Sachkov and V. E. Tarakanov,** Combinatorics of nonnegative matrices, 2002
212 **A. V. Mel′nikov, S. N. Volkov, and M. L. Nechaev,** Mathematics of financial obligations, 2002
211 **Takeo Ohsawa,** Analysis of several complex variables, 2002
210 **Toshitake Kohno,** Conformal field theory and topology, 2002
209 **Yasumasa Nishiura,** Far-from-equilibrium dynamics, 2002
208 **Yukio Matsumoto,** An introduction to Morse theory, 2002
207 **Ken'ichi Ohshika,** Discrete groups, 2002
206 **Yuji Shimizu and Kenji Ueno,** Advances in moduli theory, 2002
205 **Seiki Nishikawa,** Variational problems in geometry, 2001
204 **A. M. Vinogradov,** Cohomological analysis of partial differential equations and Secondary Calculus, 2001
203 **Te Sun Han and Kingo Kobayashi,** Mathematics of information and coding, 2002
202 **V. P. Maslov and G. A. Omel′yanov,** Geometric asymptotics for nonlinear PDE. I, 2001
201 **Shigeyuki Morita,** Geometry of differential forms, 2001
200 **V. V. Prasolov and V. M. Tikhomirov,** Geometry, 2001
199 **Shigeyuki Morita,** Geometry of characteristic classes, 2001
198 **V. A. Smirnov,** Simplicial and operad methods in algebraic topology, 2001
197 **Kenji Ueno,** Algebraic geometry 2: Sheaves and cohomology, 2001
196 **Yu. N. Lin′kov,** Asymptotic statistical methods for stochastic processes, 2001
195 **Minoru Wakimoto,** Infinite-dimensional Lie algebras, 2001
194 **Valery B. Nevzorov,** Records: Mathematical theory, 2001
193 **Toshio Nishino,** Function theory in several complex variables, 2001
192 **Yu. P. Solovyov and E. V. Troitsky,** C^*-algebras and elliptic operators in differential topology, 2001
191 **Shun-ichi Amari and Hiroshi Nagaoka,** Methods of information geometry, 2000
190 **Alexander N. Starkov,** Dynamical systems on homogeneous spaces, 2000
189 **Mitsuru Ikawa,** Hyperbolic partial differential equations and wave phenomena, 2000
188 **V. V. Buldygin and Yu. V. Kozachenko,** Metric characterization of random variables and random processes, 2000
187 **A. V. Fursikov,** Optimal control of distributed systems. Theory and applications, 2000
186 **Kazuya Kato, Nobushige Kurokawa, and Takeshi Saito,** Number theory 1: Fermat's dream, 2000

For a complete list of titles in this series, visit the
AMS Bookstore at **www.ams.org/bookstore/**.